10

APPUNTI

LECTURE NOTES

Luigi Ambrosio, Giuseppe Da Prato and Andrea Mennucci
Scuola Normale Superiore
Piazza dei Cavalieri, 7
56126 Pisa, Italy

Introduction to Measure Theory and Integration

Luigi Ambrosio, Giuseppe Da Prato
and Andrea Mennucci

Introduction to Measure Theory and Integration

EDIZIONI
DELLA
NORMALE

© 2011 Scuola Normale Superiore Pisa

ISBN 978-88-7642-385-7
ISBN 978-88-7642-386-4 (eBook)

Contents

Preface

This textbook collects the notes for an introductory course in measure theory and integration taught by the authors to undergraduate students of Scuola Normale Superiore in the last 10 years.

The goal of the course was to present, in a quick but rigorous way, the modern point of view on measure theory and integration, putting Lebesgues theory in $\mathbb{R}^n$ into a more general context and presenting the basic applications to Fourier series, calculus and real analysis. The text can also pave the way to more advanced courses in probability, stochastic processes or geometric measure theory.

Prerequisites for the book are a basic knowledge of calculus in one and several variables, metric spaces and linear algebra.

All results presented here, as well as their proofs, are classical. We claim some originality only in the presentation and in the choice of the exercises. Detailed solutions to the exercises are provided in the final part of the book.

Pisa, July 2011

Luigi Ambrosio, Giuseppe Da Prato
and Andrea Mennucci

Introduction

This course consists of an introduction to the modern theories of measure and of integration. Historically, this has been motivated by the necessity to go beyond the classical theory of Riemann's integration, usually taught in elementary Calculus courses on the real line. It is therefore useful to describe the reasons that motivate this extension.

(1) *It is not possible to give a simple, handy, characterization of the class of Riemann's integrable function, within Riemann's theory.* This is indeed possible within the stronger theory, due essentially to Lebesgue, that we are going to introduce.

(2) *The extensions of Riemann's theory to multiple integrals are very cumbersome.* This extension, useful to compute areas, volumes, etc., is known as Peano–Jordan theory, and it is sometimes taught in elementary courses of integration in more than one variable. In addition to that, important heuristic principles like Cavalieri's one can be proved only under technical and basically unnecessary regularity assumptions on the domains of integration.

(3) *Many constructive processes typical of Analysis (limits, series, integrals depending on a parameter, etc.) cannot be handled well within Riemann's theory of integration.* For instance, the following statement is true (it is a particular case of the so-called *dominated convergence theorem*):

Theorem 1. Let $f_h : [-1, 1] \to \mathbb{R}$ be continuous functions pointwise converging to a continuous function f. Assume the existence of a constant M satisfying $|f_h(x)| \leq M$ for all $x \in [-1, 1]$ and all $h \in \mathbb{N}$. Then

$$\lim_{h \to \infty} \int_{-1}^{1} f_h(x)\, dx = \int_{-1}^{1} f(x)\, dx.$$

Even though this statement makes perfectly sense within Riemann's theory, any attempt to prove this result within the theory (try, if you don't believe!) seems to fail, and leads (more or less explicitely, see [2]) to a larger theory. In addition to that, the continuity assumption on the limit function f is not natural, because a pointwise limit of continuous functions need not be continuous, and we would like to give a sense to $\int_{-1}^{1} f(x)\,dx$ even without this assumption. This necessity emerges for instance in the study of the convergence of Fourier series

$$f(x) = \sum_{i=0}^{\infty} a_i \cos(i\pi x) + b_i \sin(i\pi x) \qquad x \in [-1, 1].$$

In this case the uniform convergence of the series, which implies the continuity of f as well, is ensured by the condition $\sum_i |a_i| + |b_i| < \infty$. On the other hand, we will see that the "natural" condition for the convergence (in a suitable sense) of the series is much weaker:

$$\sum_{i=0}^{\infty} a_i^2 + b_i^2 < \infty.$$

Under this condition the limit function f need not be continuous: for instance, if $f(x) = 1$ for $x \in [-1/2, 1/2]$ and $f(x) = 0$ otherwise, then we will see that the coefficients of the Fourier series are given by $b_i = 0$ for all i (because f is even) and by

$$a_i = \begin{cases} \dfrac{1}{2} & \text{if } i = 0; \\[2ex] \dfrac{\sin(\pi i/2)}{i\pi} & \text{if } i > 0. \end{cases}$$

(4) *The spaces of integrable functions, as for instance*

$$H := \left\{ f : [-1, 1] \to \mathbb{R} : \int_{-1}^{1} f^2(x)\,dx < \infty \right\}$$

endowed with the scalar product

$$\langle f, g \rangle := \int_{-1}^{1} f(x)g(x)\,dx$$

and with the (pseudo) induced distance $d(f, g) = \langle f - g, f - g \rangle^{1/2}$, *are not complete, if we restrict ourselves to Riemann integrable functions only.* In this sense, the path from Riemann's to Lebesgue's theory is the same one that led from the (incomplete) set of rational numbers $\mathbb{Q}$ to the (complete) real line $\mathbb{R}$.

Lebesgue's theory extends Riemann's theory in two independent directions. The first one is concerned, as we already said, with more general classes of functions, not necessarily continuous or piecewise continuous (the so-called *Borel* or *measurable* functions). The second direction can be better understood if we remind the very definition of Riemann's integral

$$\int_{-1}^{1} f(x)\, dx \sim \sum_{i=1}^{n-1} (t_{i+1} - t_i) f(t_i)$$

where $t_1 = -1$, $t_n = 1$ and the approximation is better and better as the parameter $\sup_{i<n} t_{i+1} - t_i$ tends to 0. More generally, instead of integrating with respect to the "length" measure, we can integrate with respect to a generic measure μ and define

$$\int_{-1}^{1} f(x)\, d\mu(x) \sim \sum_{i=1}^{n} \mu(A_i) f(x_i) \tag{1}$$

where $A_1, \ldots, A_n$ is a partition of $[-1, 1]$, $x_i \in A_i$ and the approximation is expected to be better and better as the parameter $\sup_i \operatorname{diam}(A_i)$ tends to 0. We may think, for instance to $[-1, 1]$ as a possibly non-homogeneous bar, and to $\mu(A)$ as the "mass" of the subset A of the bar: because of non-homogeneity, $\mu(A)$ need not be proportional to the length of A.

Once we adopt this viewpoint, we will see that it is not hard to obtain a theory of integration in general metric spaces, and even in more general classes of spaces. On the other hand, the approximation (1), that in any case clarifies the intuitive meaning of the integral, will remain valid for continuous functions only.

Chapter 1
Measure spaces

In this chapter we shall introduce all basic concepts of measure theory, adopting the point of view of measures as set functions. The domains of measures may have different stability properties, and this leads to the concepts of ring, algebra and σ–algebra. The most basic tool developed in the chapter is Carathéodory's theorem, which ensures in many cases the existence and the uniqueness of a σ–additive measure having some prescribed values on a set of generators of the σ–algebra. In the final part of the chapter we will apply these abstract tools to the problem of constructing a "length" measure on the real line, the so-called *Lebesgue measure*, and we will study its main properties.

1.1. Notation and preliminaries

We denote by $\mathbb{N} = \{0, 1, 2, \ldots\}$ the set of natural numbers, and by $\mathbb{N}^*$ the set of positive natural numbers. Unless otherwise stated, sequences will always be indexed by natural numbers.

We shall denote by X a non-empty set, by $\mathscr{P}(X)$ the set of all parts of X and by $\varnothing$ the empty set. For any subset A of X we shall denote by A^c its complement $A^c := \{x \in X : x \notin A\}$. If $A, B \in \mathscr{P}(X)$ we denote by $A \setminus B$ the *relative complement* $A \cap B^c$, and by $A \triangle B$ the *symmetric difference* $(A \setminus B) \cup (B \setminus A)$.

Let (A_n) be a sequence in $\mathscr{P}(X)$. Then the following *De Morgan* identity holds,

$$\left(\bigcup_{n=0}^{\infty} A_n \right)^c = \bigcap_{n=0}^{\infty} A_n^c.$$

Moreover, we define [1]

$$\limsup_{n \to \infty} A_n := \bigcap_{n=0}^{\infty} \bigcup_{m=n}^{\infty} A_m, \qquad \liminf_{n \to \infty} A_n := \bigcup_{n=0}^{\infty} \bigcap_{m=n}^{\infty} A_m.$$

---

[1] Notice the analogy with liminf and limsup limits for a sequence (a_n) of real numbers. We have $\limsup_{n \to \infty} a_n = \inf_{n \in \mathbb{N}} \sup_{m \geq n} a_m$ and $\liminf_{n \to \infty} a_n = \sup_{n \in \mathbb{N}} \inf_{m \geq n} a_m$. This is something more than an analogy, see Exercise 1.1.

As it can be easily checked, $\limsup_n A_n$ (respectively $\liminf_n A_n$) consists of those elements of X that belong to infinitely many A_n (respectively that belong to all but finitely many A_n).

It easy to check that if (A_n) is nondecreasing (*i.e.* $A_n \subset A_{n+1}$, $n \in \mathbb{N}$), we have

$$\liminf_{n \to \infty} A_n = \limsup_{n \to \infty} A_n = \bigcup_{n=0}^{\infty} A_n,$$

whereas if (A_n) is nonincreasing (*i.e.* $A_n \supset A_{n+1}$, $n \in \mathbb{N}$), we have

$$\liminf_{n \to \infty} A_n = \limsup_{n \to \infty} A_n = \bigcap_{n=0}^{\infty} A_n.$$

In the first case we shall write $A_n \uparrow L$, and in the second one $A_n \downarrow L$.

1.2. Rings, algebras and σ–algebras

Definition 1.1 (Rings and Algebras). A non empty subset $\mathscr{A}$ of $\mathscr{P}(X)$ is said to be a *ring* if:

  (i) $\varnothing$ belongs to $\mathscr{A}$;
  (ii) $A, B \in \mathscr{A} \implies A \cup B, A \cap B \in \mathscr{A}$;
  (iii) $A, B \in \mathscr{A} \implies A \setminus B \in \mathscr{A}$.

We say that a ring is an *algebra* if $X \in \mathscr{A}$.

Notice that rings are stable only with respect to relative complement, whereas algebras are stable under complement in X.

Let $\mathscr{K} \subset \mathscr{P}(X)$. As the intersection of any family of algebras is still an algebra, the minimal algebra including $\mathscr{K}$ (that is the intersection of all algebras including $\mathscr{K}$) is well defined, and called the *algebra generated by* $\mathscr{K}$. A constructive characterization of the algebra generated by $\mathscr{K}$ can be easily achieved as follows: set $\mathscr{F}^{(0)} = \mathscr{K} \cup \{\varnothing\}$ and

$$\mathscr{F}^{(i+1)} := \bigcup \left\{ A \cup B,\ A^c\ :\ A, B \in \mathscr{F}^{(i)} \right\} \qquad \forall i \geq 0.$$

Then, the algebra $\mathscr{A}$ generated by $\mathscr{K}$ is given by $\bigcup_i \mathscr{F}^{(i)}$. Indeed, it is immediate to check by induction on i that $\mathscr{A} \supset \mathscr{F}^{(i)}$, and therefore the union of the $\mathscr{F}^{(i)}$'s is contained in $\mathscr{A}$. On the other hand, this union is easily seen to be an algebra, so the minimality of $\mathscr{A}$ provides the opposite inclusion.

Definition 1.2 (σ–algebras). A non-empty subset $\mathscr{E}$ of $\mathscr{P}(X)$ is said to be a σ–*algebra* if:

  (i) $\mathscr{E}$ is an algebra;

(ii) if (A_n) is a sequence of elements of $\mathscr{E}$ then $\bigcup\limits_{n=0}^{\infty} A_n \in \mathscr{E}$.

If $\mathscr{E}$ is a σ–algebra and $(A_n) \subset \mathscr{E}$ we have $\bigcap_n A_n \in \mathscr{E}$ by the De Morgan identity. Moreover, both sets

$$\liminf_{n\to\infty} A_n, \qquad \limsup_{n\to\infty} A_n,$$

belong to $\mathscr{E}$.

Obviously, $\{\varnothing, X\}$ and $\mathscr{P}(X)$ are σ–algebras, respectively the smallest and the largest ones. Let $\mathscr{K}$ be a subset of $\mathscr{P}(X)$. As the intersection of any family of σ–algebras is still a σ-algebra, the minimal σ–algebra including $\mathscr{K}$ (that is the intersection of all σ–algebras including $\mathscr{K}$) is well defined, and called the *σ–algebra generated by $\mathscr{K}$*. It is denoted by $\sigma(\mathscr{K})$.

In contrast with the case of generated algebras, it is quite hard to give a constructive characterization of the generated σ-algebras: this requires the transfinite induction and it is illustrated in Exercise 1.18.

Definition 1.3 (Borel σ-algebra). If (E, d) is a metric space, the σ–algebra generated by all open subsets of E is called the *Borel σ–algebra* of E and it is denoted by $\mathscr{B}(E)$.

In the case when $E = \mathbb{R}$ the Borel σ-algebra has a particularly simple class of generators.

Example 1.4 ($\mathscr{B}(\mathbb{R})$). Let $\mathscr{I}$ be the set of all semi–closed intervals $[a, b)$ with $a \le b$. Then $\sigma(\mathscr{I})$ coincides with $\mathscr{B}(\mathbb{R})$. In fact $\sigma(\mathscr{I})$ contains all open intervals (a, b) since

$$(a, b) = \bigcup_{n=n_0}^{\infty} \left[a + \frac{1}{n}, b\right) \qquad \text{with } n_0 > \frac{1}{b - a}.$$

Moreover, any open set A in $\mathbb{R}$ is a countable union of open intervals.[2] An analogous argument proves that $\mathscr{B}(\mathbb{R})$ is generated by semi-closed intervals $(a, b]$, by open intervals, by closed intervals and even by open or closed half-lines.

---

[2] Indeed, let (a_k) be a sequence including all rational numbers of A and denote by I_k the largest open interval contained in A and containing a_k. We clearly have $A \supset \bigcup_{k=0}^{\infty} I_k$, but also the opposite inclusion holds: it suffices to consider, for any $x \in A$, $r > 0$ such that $(x - r, x + r) \subset A$, and k such that $a_k \in (x - r, x + r)$ to obtain $(x - r, x + r) \subset I_k$, by the maximality of I_k, and then $x \in I_k$.

1.3. Additive and σ–additive functions

Let $\mathscr{A} \subset \mathscr{P}(X)$ be a ring and let μ be a mapping from $\mathscr{A}$ into $[0, +\infty]$ such that $\mu(\varnothing) = 0$. We say that μ is *additive* if

$$A, \; B \in \mathscr{A}, \;\; A \cap B = \varnothing \quad \Longrightarrow \quad \mu(A \cup B) = \mu(A) + \mu(B).$$

If μ is additive, $A, \; B \in \mathscr{A}$ and $A \supset B$, we have $\mu(A) = \mu(B) + \mu(A \setminus B)$, so that $\mu(A) \geq \mu(B)$. Therefore any additive function is *nondecreasing* with respect to set inclusion. Moreover, by applying repeatedly the additivity property, additive measures satisfy

$$\mu\left(\bigcup_{k=1}^{n} A_k\right) = \sum_{k=1}^{n} \mu(A_k)$$

for $n \in \mathbb{N}^*$ and mutually disjoint sets $A_1, \ldots, A_n \in \mathscr{A}$.

A set function μ on $\mathscr{A}$ is called σ–*additive* if $\mu(\varnothing) = 0$ and for any sequence $(A_n) \subset \mathscr{A}$ of mutually disjoint sets such that $\bigcup_n A_n \in \mathscr{A}$ we have

$$\mu\left(\bigcup_{n=0}^{\infty} A_n\right) = \sum_{n=0}^{\infty} \mu(A_n).$$

Obviously σ–additive functions are additive, because we can consider countable unions in which only finitely many A_n are nonempty.

Another useful concept is the σ–*subadditivity*: we say that μ is σ–subadditive if

$$\mu(B) \leq \sum_{n=0}^{\infty} \mu(A_n),$$

for any $B \in \mathscr{A}$ and any sequence $(A_n) \subset \mathscr{A}$ such that $B \subset \bigcup_n A_n$. Notice that, unlike the definition of σ–additivity, the sets A_n need not be disjoint here.

Remark 1.5 (σ–additivity and σ–subadditivity). Let μ be additive on a ring $\mathscr{A}$ and let $(A_n) \subset \mathscr{A}$ be mutually disjoint and such that $\bigcup_n A_n \in \mathscr{A}$. Then by monotonicity we have

$$\mu\left(\bigcup_{n=0}^{\infty} A_n\right) \geq \mu\left(\bigcup_{n=0}^{k} A_n\right) = \sum_{n=0}^{k} \mu(A_n), \quad \text{for all } k \in \mathbb{N}.$$

Therefore, letting $k \uparrow \infty$ we get

$$\mu\left(\bigcup_{n=0}^{\infty} A_n\right) \geq \sum_{n=0}^{\infty} \mu(A_n).$$

Thus, to show that an additive function is σ–additive, it is enough to prove that it is σ–subadditive.

Conversely, it is not difficult to show that σ–additive set functions are σ–subadditive: indeed, if $B \subset \bigcup_n A_n$ we can define $A'_0 = B \cap A_0$ and $A'_n := B \cap A_n \setminus \bigcup_{m<n} A_m$ for $n \in \mathbb{N}^*$, so that B is the disjoint union of the sets A'_n, to obtain

$$\mu(B) = \sum_{n=0}^{\infty} \mu(A'_n) \le \sum_{n=0}^{\infty} \mu(A_n).$$

Let μ be additive on $\mathscr{A}$. Then σ–additivity of μ is equivalent to continuity of μ in the sense of the following proposition.

Proposition 1.6 (Continuity on nondecreasing sequences). *If μ is additive on a ring $\mathscr{A}$, then (i) $\Longleftrightarrow$ (ii), where:*

(i) *μ is σ–additive;*
(ii) *$(A_n) \subset \mathscr{A}$ and $A \in \mathscr{A}$, $A_n \uparrow A \Longrightarrow \mu(A_n) \uparrow \mu(A)$.*

Proof. (i)$\Longrightarrow$(ii). In the proof of this implication we can assume with no loss of generality that $\mu(A_n) < \infty$ for all $n \in \mathbb{N}$. Let $(A_n) \subset \mathscr{A}$, $A \in \mathscr{A}$, $A_n \uparrow A$. Then

$$A = A_0 \cup \bigcup_{n=0}^{\infty}(A_{n+1} \setminus A_n),$$

the unions being disjoint. Since μ is σ–additive, we deduce that

$$\mu(A) = \mu(A_0) + \sum_{n=0}^{\infty}(\mu(A_{n+1}) - \mu(A_n)) = \lim_{n \to \infty} \mu(A_n),$$

and (ii) follows.

(ii)$\Longrightarrow$(i). Let $(A_n) \subset \mathscr{A}$ be mutually disjoint and such that $A := \bigcup_n A_n \in \mathscr{A}$. Set

$$B_m := \bigcup_{k=0}^{m} A_k.$$

Then $B_m \uparrow A$ and $\mu(B_m) = \sum_{0}^{m} \mu(A_k) \uparrow \mu(A)$ by the assumption. This implies (i). $\qquad\square$

Proposition 1.7 (Continuity on nonincreasing sequences). *Let μ be σ–additive on a ring $\mathscr{A}$. Then*

$$(A_n) \subset \mathscr{A} \text{ and } A \in \mathscr{A}, A_n \downarrow A, \mu(A_0) < \infty \implies \mu(A_n) \downarrow \mu(A).$$

$$(1.1)$$

Proof. Setting $B_n := A_0 \setminus A_n$, $B := A_0 \setminus A$, we have $B_n \uparrow B$, therefore the previous proposition gives $\mu(B_n) \uparrow \mu(B)$. As $\mu(A_n) = \mu(A_0) - \mu(B_n)$ and $\mu(A) = \mu(A_0) - \mu(B)$ the proof is achieved. $\qquad\square$

Corollary 1.8 (Upper and lower semicontinuity of the measure). *Let μ be σ–additive on a σ–algebra $\mathcal{E}$ and let $(A_n) \subset \mathcal{E}$. Then we have*

$$\mu\left(\liminf_{n\to\infty} A_n\right) \le \liminf_{n\to\infty} \mu(A_n) \tag{1.2}$$

and, if $\mu(X) < \infty$, we have also

$$\limsup_{n\to\infty} \mu(A_n) \le \mu\left(\limsup_{n\to\infty} A_n\right). \tag{1.3}$$

Proof. Set $L := \limsup_n A_n$. Then we can write

$$L = \bigcap_{n=0}^{\infty} B_n \qquad \text{where} \qquad B_n := \bigcup_{m=n}^{\infty} A_m. \tag{1.4}$$

Now, assuming $\mu(X) < \infty$, by Proposition 1.7 it follows that

$$\mu(L) = \lim_{n\to\infty} \mu(B_n) = \inf_{n\in\mathbb{N}} \mu(B_n) \ge \inf_{n\in\mathbb{N}} \sup_{m\ge n} \mu(A_m) = \limsup_{n\to\infty} \mu(A_n).$$

Thus, we have proved (1.3). The inequality (1.2) can be proved similarly using Proposition 1.6, thus without using the assumption $\mu(X) < \infty$. $\quad\square$

The following result is very useful to estimate the measure of a lim sup of sets.

Lemma 1.9. *Let μ be σ–additive on a σ–algebra $\mathcal{E}$ and let $(A_n) \subset \mathcal{E}$. Assume that $\sum_{n=0}^{\infty} \mu(A_n) < \infty$. Then*

$$\mu\left(\limsup_{n\to\infty} A_n\right) = 0.$$

Proof. Set $L = \limsup_{n\to\infty} A_n$ and define B_n as in (1.4). Then the inclusion $L \subset B_n$ gives

$$\mu(L) \le \mu(B_n) \le \sum_{m=n}^{\infty} \mu(A_m) \qquad \text{for all } n \in \mathbb{N}.$$

As $n \to \infty$ we find $\mu(L) = 0$. $\qquad\square$

1.4. Measurable spaces and measure spaces

Let $\mathscr{E}$ be a σ–algebra of subsets of X. Then we say that the pair $(X, \mathscr{E})$ is a *measurable space*. Let $\mu : \mathscr{E} \to [0, +\infty]$ be a σ–additive function. Then we call μ a *measure* on $(X, \mathscr{E})$, and we call the triple $(X, \mathscr{E}, \mu)$ a *measure* space.

The measure μ is said to be *finite* if $\mu(X) < \infty$, σ–*finite* if there exists a sequence $(A_n) \subset \mathscr{E}$ such that $\bigcup_n A_n = X$ and $\mu(A_n) < \infty$ for all $n \in \mathbb{N}$. Finally, μ is called a *probability measure* if $\mu(X) = 1$.

The simplest (but fundamental) example of a probability measure is the *Dirac mass* δ_x, defined by

$$
\delta_x(B) := \begin{cases} 1 & \text{if } x \in B \\ 0 & \text{if } x \notin B. \end{cases}
$$

This example can be generalized as follows, see also Exercise 1.5 and Exercise 1.23.

Example 1.10 (Discrete measures). Assume that $Y \subset X$ is a finite or countable set. Given $c : Y \to [0, +\infty]$ we can define a measure on $(X, \mathscr{P}(X))$ as follows:

$$
\mu(B) := \sum_{x \in B \cap Y} c(x) \qquad \forall B \subset X.
$$

Clearly $\mu = \sum_{x \in Y} c(x)\delta_x$ is a finite measure if and only if $\sum_{x \in Y} c(x) < \infty$, and it is σ–finite if and only if $c(x) \in [0, +\infty)$ for all $x \in Y$.

More generally, the construction above works even when Y is uncountable, by replacing the sum with

$$
\sup_{c \in B \cap Y'} \sum c(x),
$$

where the supremum is made among the finite subsets Y' of Y. The measures arising in the previous example are called *atomic*, and clearly if X is either finite or countable then any measure μ in $(X, \mathscr{P}(X))$ is atomic: it suffices to notice that

$$
\mu = \sum_{x \in X} c(x)\delta_x \qquad \text{with} \qquad c(x) := \mu(\{x\}).
$$

In the next section we will introduce a fundamental tool for the construction of non-atomic measures.

Definition 1.11 (μ–negligible sets and μ–almost everywhere). Given a measure space $(X, \mathcal{E}, \mu)$, we say that $B \in \mathcal{E}$ is μ–*negligible* if $\mu(B) = 0$, and we say that a property $P(x)$ holds μ–almost everywhere if the set

$$\{x \in X \ : \ P(x) \text{ is false}\}$$

is contained in a μ–negligible set.

Notice that the class of μ–negligible sets is stable under finite or countable unions. It is sometimes convenient to know that any subset of a μ–negligible set is still μ–negligible.

Definition 1.12 (μ–completion of a σ–algebra and μ–measurable sets). Let $(X, \mathcal{E}, \mu)$ be a measure space. We define

$$\mathcal{E}_\mu := \{A \in \mathscr{P}(X) \ : \ \text{for some } B, C \in \mathcal{E} \text{ with } \mu(C) = 0, A \Delta B \subset C\}.$$

It is easy to check that $\mathcal{E}_\mu$ is still a σ–algebra, the so-called *completion* of $\mathcal{E}$ with respect to μ. The elements of $\mathcal{E}_\mu$ are called μ–measurable sets.

It is also easy to check that μ can be extended to all $A \in \mathcal{E}_\mu$ simply by setting $\mu(A) = \mu(B)$, where $B \in \mathcal{E}$ is any set such that $A \Delta B$ is contained in a μ–negligible set of $\mathcal{E}$. This extension is well defined (*i.e.* independent of the choice of B), still σ–additive and μ–negligible sets coincide with those sets that are contained in some $B \in \mathcal{E}$ with $\mu(B) = 0$. As a consequence, any subset of a μ–negligible set is μ–negligible as well.

1.5. The basic extension theorem

The following result, due to Carathéodory, allows to extend a σ–additive function on a ring $\mathscr{A}$ to a σ–additive function on $\sigma(\mathscr{A})$. It is one of the basic tools in the construction of non-trivial measures in many cases of interest, as we will see.

Theorem 1.13 (Carathéodory). *Let $\mathscr{A} \subset \mathscr{P}(X)$ be a ring, and let $\mathcal{E}$ be the σ–algebra generated by $\mathscr{A}$. Let $\mu \colon \mathscr{A} \to [0, +\infty]$ be σ–additive. Then μ can be extended to a measure on $\mathcal{E}$. If μ is σ–finite, i.e. there exist $A_n \in \mathscr{A}$ with $A_n \uparrow X$ and $\mu(A_n) < \infty$ for any n, then the extension is unique.*

To prove this theorem we need some preliminaries: for the uniqueness the *Dynkin theorem* and for the existence the concepts of *outer measure* and *additive set*.

1.5.1. Dynkin systems

A non-empty subset $\mathcal{K}$ of $\mathcal{P}(X)$ is called a π-*system* if

$$A,\, B \in \mathcal{K} \Longrightarrow A \cap B \in \mathcal{K}.$$

A non-empty subset $\mathcal{D}$ of $\mathcal{P}(X)$ is called a *Dynkin system* if

(i) $X,\, \varnothing \in \mathcal{D}$;
(ii) $A \in \mathcal{D} \Longrightarrow A^c \in \mathcal{D}$;
(iii) $(A_i) \subset \mathcal{D}$ mutually disjoint $\Longrightarrow \bigcup_i A_i \in \mathcal{D}$.

Obviously any σ-algebra is a Dynkin system. Moreover, if $\mathcal{D}$ is both a Dynkin system and a π-system then it is a σ-algebra. In fact, if (A_i) is a sequence in $\mathcal{D}$ of not necessarily disjoint sets we have

$$\bigcup_{i=0}^{\infty} A_i = A_0 \cup (A_1 \setminus A_0) \cup ((A_2 \setminus A_1) \setminus A_0) \cup \cdots$$

and so $\bigcup_i A_i \in \mathcal{D}$ by (ii) and (iii).

Let us prove now the following important result.

Theorem 1.14 (Dynkin). *Let $\mathcal{K}$ be a π-system and let $\mathcal{D} \supset \mathcal{K}$ be a Dynkin system. Then $\sigma(\mathcal{K}) \subset \mathcal{D}$.*

Proof. Let $\mathcal{D}_0$ be the minimal Dynkin system including $\mathcal{K}$. We are going to show that $\mathcal{D}_0$ is a σ-algebra which will prove the theorem. For this it is enough to show, as remarked before, that the following implication holds:

$$A,\, B \in \mathcal{D}_0 \Longrightarrow A \cap B \in \mathcal{D}_0. \tag{1.5}$$

For any $B \in \mathcal{D}_0$ we set

$$\mathcal{H}(B) = \{F \in \mathcal{D}_0 \; : \; B \cap F \in \mathcal{D}_0\}.$$

We claim that $\mathcal{H}(B)$ is a Dynkin system. In fact properties (i) and (iii) are clear. It remains to show that if $F \cap B \in \mathcal{D}_0$ then $F^c \cap B \in \mathcal{D}_0$ or, equivalently, $F \cup B^c \in \mathcal{D}_0$. In fact, since $F \cup B^c = (F \setminus B^c) \cup B^c = (F \cap B) \cup B^c$ and $F \cap B$ and B^c are disjoint, we have that $F \cup B^c \in \mathcal{D}_0$ as required.

Notice first that if $K \in \mathcal{K}$ we have $\mathcal{K} \subset \mathcal{H}(K)$ since $\mathcal{K}$ is a π-system. Therefore $\mathcal{H}(K) = \mathcal{D}_0$, by the minimality of $\mathcal{D}_0$. Consequently, the following implication holds

$$K \in \mathcal{K},\, B \in \mathcal{D}_0 \Longrightarrow K \cap B \in \mathcal{D}_0,$$

which implies $\mathcal{K} \subset \mathcal{H}(B)$ for all $B \in \mathcal{D}_0$. Again, the fact that $\mathcal{H}(B)$ is a Dynkin system and the minimality of $\mathcal{D}_0$ give that $\mathcal{H}(B) = \mathcal{D}_0$ for all $B \in \mathcal{D}_0$. By the definition of $\mathcal{H}(B)$, this proves (1.5). $\qquad\square$

The uniqueness part in Caratheodory's theorem is a direct consequence of the following coincidence criterion for measures; in turn, the proof of the criterion relies on Theorem 1.14.

Proposition 1.15 (Coincidence criterion). *Let μ_1, μ_2 be measures in $(X, \mathcal{E})$ and assume that:*

(i) *the coincidence set*

$$\mathcal{D} := \{A \in \mathcal{E} \ : \ \mu_1(A) = \mu_2(A)\}$$

contains a π–system $\mathcal{K}$ with $\sigma(\mathcal{K}) = \mathcal{E}$;
(ii) *there exists a nondecreasing sequence $(X_i) \subset \mathcal{K}$ with $\mu_1(X_i) = \mu_2(X_i) < \infty$ and $X_i \uparrow X$.*

Then $\mu_1 = \mu_2$.

Proof. We first assume that $\mu_1(X) = \mu_2(X)$ is finite. Under this assumption $\mathcal{D}$ is a Dynkin system including the π–system $\mathcal{K}$ (stability of $\mathcal{D}$ under complement is ensured precisely by the finiteness assumption). Thus, by the Dynkin theorem, $\mathcal{D} = \mathcal{E}$, which implies that $\mu_1 = \mu_2$.

Assume now that we are in the general case and let X_i be given by assumption (ii). Fix $i \in \mathbb{N}$ and define the σ–algebra $\mathcal{E}_i$ of subsets of X_i by

$$\mathcal{E}_i := \{A \subset X_i \ : \ A \in \mathcal{E}\}.$$

We may obviously consider μ_1 and μ_2 as finite measures in the measurable space $(X_i, \mathcal{E}_i)$. Since these measures coincide on the π–system

$$\mathcal{K}_i := \{A \subset X_i \ : \ A \in \mathcal{K}\}$$

we obtain, by the previous step, that μ_1 and μ_2 coincide on $\sigma(\mathcal{K}_i) \subset \mathcal{P}(X_i)$.

Now, let us prove the inclusion

$$\{B \in \mathcal{E} \ : \ B \subset X_i\} \subset \sigma(\mathcal{K}_i). \tag{1.6}$$

Indeed

$$\big\{B \subset X \ : \ B \cap X_i \in \sigma(\mathcal{K}_i)\big\}$$

is a σ–algebra containing $\mathcal{K}$ (here we use the fact that $X_i \in \mathcal{K}$), and therefore contains $\mathcal{E}$. Hence any element of $\mathcal{E}$ contained in X_i belongs to $\sigma(\mathcal{K}_i)$.

By (1.6) we obtain $\mu_1(B \cap X_i) = \mu_2(B \cap X_i)$ for all $B \in \mathcal{E}$ and all $i \in \mathbb{N}$. Passing to the limit as $i \to \infty$, since B is arbitrary we obtain that $\mu_1 = \mu_2$. $\qquad\square$

1.5.2. The outer measure

Let μ be a set function defined on $\mathscr{A} \subset \mathscr{P}(X)$. For any $E \in \mathscr{P}(X)$ we define:

$$\mu^*(E) := \inf\left\{ \sum_{i=0}^{\infty} \mu(A_i) \ : \ A_i \in \mathscr{A}, \ E \subset \bigcup_{i=0}^{\infty} A_i \right\}.$$

μ^* is called the *outer* measure induced by μ. We can easily show that μ^* is a nondecreasing set function, namely $\mu^*(E) \leq \mu^*(F)$ whenever $E \subset F \subset X$.

We will obtain the proof of the existence part of Carathéodory's theorem by showing in the proposition below that μ^* extends μ if μ is $\sigma-$subadditive, and that (Theorem 1.17) μ^* is σ-additive on a σ-algebra containing $\sigma(\mathscr{A})$ if μ is $\mathscr{A}$ is a ring and μ is additive on $\mathscr{A}$. In particular if μ is $\sigma-$additive on $\mathscr{A}$ we see that μ^* provides the desired $\sigma-$additive extension to $\sigma(\mathscr{A})$.

Proposition 1.16. *The set function μ^* is $\sigma-$subadditive on $\mathscr{P}(X)$ and extends μ if μ is $\sigma-$subadditive on $\mathscr{A}$ and $\mu(\varnothing) = 0$.*

Proof. Let $(E_i) \subset \mathscr{P}(X)$ and set $E := \bigcup_i E_i$. Assume that $\sum_i \mu^*(E_i)$ are finite (otherwise the assertion is trivial). Then, since $\mu^*(E_i)$ is finite for any $i \in \mathbb{N}$, for any $\varepsilon > 0$ there exist $A_{i,j} \in \mathscr{A}$ such that

$$\sum_{j=0}^{\infty} \mu(A_{i,j}) < \mu^*(E_i) + \frac{\varepsilon}{2^{i+1}}, \qquad E_i \subset \bigcup_{j=0}^{\infty} A_{i,j}, \qquad i \in \mathbb{N}.$$

Consequently

$$\sum_{i,\,j=0}^{\infty} \mu(A_{i,j}) \leq \sum_{i=0}^{\infty} \mu^*(E_i) + \varepsilon.$$

Since $E \subset \bigcup_{i,j=0}^{\infty} A_{i,j}$ we have

$$\mu^*(E) \leq \sum_{i,j=0}^{\infty} \mu(A_{i,j}) \leq \sum_{i=0}^{\infty} \mu^*(E_i) + \varepsilon$$

and the first part of the statement follows from the arbitrariness of ε.

Now, let us assume that μ is σ-subadditive on $\mathscr{A}$ and choose $E \in \mathscr{A}$; since $E \subset \bigcup_i A_i$ then $\mu(E) \leq \sum_i \mu(A_i)$, so we deduce $\mu^*(E) \geq \mu(E)$; but, by choosing $A_0 = E$ and $A_n = \varnothing$ for $n \geq 1$, we obtain that $\mu^*(E) = \mu(E)$. This proves that μ^* extends μ. $\qquad\qquad\square$

Let us now define the *additive sets*, according to Carathéodory. A set $A \in \mathscr{P}(X)$ is called *additive* if

$$\mu^*(E) = \mu^*(E \cap A) + \mu^*(E \cap A^c) \quad \text{for all } E \in \mathscr{P}(X). \qquad (1.7)$$

We denote by $\mathscr{G}$ the family of all additive sets.

Notice that, since μ^* is subadditive, (1.7) is equivalent to

$$\mu^*(E) \geq \mu^*(E \cap A) + \mu^*(E \cap A^c) \quad \text{for all } E \in \mathscr{P}(X). \qquad (1.8)$$

Obviously, the class $\mathscr{G}$ of additive sets is stable under complement; moreover, by taking $E = A \cup B$ with $A \in \mathscr{G}$ and $A \cap B = \emptyset$, we obtain the additivity property

$$\mu^*(A \cup B) = \mu^*(A) + \mu^*(B). \qquad (1.9)$$

Other important properties of $\mathscr{G}$ are listed in the next proposition.

Theorem 1.17. *Assume that $\mathscr{A}$ is a ring and that μ is additive. Then $\mathscr{G}$ is a σ–algebra containing $\mathscr{A}$ and μ^* is σ–additive on $\mathscr{G}$.*

Proof. We proceed in three steps: we show that $\mathscr{G}$ contains $\mathscr{A}$, that $\mathscr{G}$ is a σ–algebra and that μ^* is additive on $\mathscr{G}$. As pointed in Remark 1.5, if μ^* is σ–subadditive and additive on the σ–algebra $\mathscr{G}$, then μ^* is σ–additive.
Step 1. $\mathscr{A} \subset \mathscr{G}$. Let $A \in \mathscr{A}$ and $E \in \mathscr{P}(X)$, we have to show (1.8). Assume $\mu^*(E) < \infty$ (otherwise (1.8) trivially holds), fix $\varepsilon > 0$ and choose $(B_i) \subset \mathscr{A}$ such that

$$E \subset \bigcup_{i=0}^{\infty} B_i, \quad \mu^*(E) + \varepsilon > \sum_{i=0}^{\infty} \mu(B_i).$$

Then, by the definition of μ^*, it follows that

$$\mu^*(E) + \varepsilon > \sum_{i=0}^{\infty} \mu(B_i) = \sum_{i=0}^{\infty} [\mu(B_i \cap A) + \mu(B_i \cap A^c)]$$
$$\geq \mu^*(E \cap A) + \mu^*(E \cap A^c).$$

Since ε is arbitrary we have $\mu^*(E) \geq \mu^*(E \cap A) + \mu^*(E \cap A^c)$, and (1.8) follows.
Step 2. $\mathscr{G}$ is an algebra and μ^* is additive on $\mathscr{G}$. We already know that $A \in \mathscr{G}$ implies $A^c \in \mathscr{G}$. Let us prove now that if $A, B \in \mathscr{G}$ then $A \cup B \in \mathscr{G}$. For any $E \in \mathscr{P}(X)$ we have

$$\mu^*(E) = \mu^*(E \cap A) + \mu^*(E \cap A^c)$$

$$= \mu^*(E \cap A) + \mu^*(E \cap A^c \cap B) + \mu^*(E \cap A^c \cap B^c) \qquad (1.10)$$

$$= [\mu^*(E \cap A) + \mu^*(E \cap A^c \cap B)] + \mu^*(E \cap (A \cup B)^c).$$

Since

$$(E \cap A) \cup (E \cap A^c \cap B) = E \cap (A \cup B),$$

we have by the subadditivity of μ^*,

$$\mu^*(E \cap A) + \mu^*(E \cap A^c \cap B) \geq \mu^*(E \cap (A \cup B)).$$

So, by (1.10) it follows that

$$\mu^*(E) \geq \mu^*(E \cap (A \cup B)) + \mu^*(E \cap (A \cup B)^c),$$

and $A \cup B \in \mathscr{G}$ as required. The additivity of μ^* on $\mathscr{G}$ follows directly from (1.9).

Step 3. $\mathscr{G}$ is a σ–algebra. Let $(A_n) \subset \mathscr{G}$. We are going to show that $S := \bigcup_n A_n \in \mathscr{G}$. Since we know that $\mathscr{G}$ is an algebra, it is not restrictive to assume that all sets A_n are mutually disjoint. Set $S_n := \bigcup_0^n A_i$, for $n \in \mathbb{N}$.

For any $n \in \mathbb{N}$, by using the σ–subadditivity of μ^* and by applying (1.7) repeatedly, we get

$$\mu^*(E \cap S^c) + \mu^*(E \cap S) \leq \mu^*(E \cap S^c) + \sum_{i=0}^{\infty} \mu^*(E \cap A_i)$$

$$= \lim_{n \to \infty} \left[\mu^*(E \cap S^c) + \sum_{i=0}^{n} \mu^*(E \cap A_i) \right]$$

$$= \lim_{n \to \infty} \left[\mu^*(E \cap S^c) + \mu^*(E \cap S_n) \right].$$

Since $S^c \subset S_n^c$ it follows that

$$\mu^*(E \cap S^c) + \mu^*(E \cap S) \leq \limsup_{n \to \infty} \left[\mu^*(E \cap S_n) + \mu^*(E \cap S_n^c) \right]$$

$$= \mu^*(E).$$

So, $S \in \mathscr{G}$ and $\mathscr{G}$ is a σ–algebra. $\qquad\square$

Remark 1.18. We have proved that

$$\sigma(\mathscr{A}) \subset \mathscr{G} \subset \mathscr{P}(X). \tag{1.11}$$

One can show that the inclusions above are strict in general, for instance when μ is the Lebesgue measure we shall consider in the next section. In fact, in the case when $X = \mathbb{R}$ and $\sigma(\mathscr{A})$ is the Borel σ-algebra, Exercise 1.19 shows that $\sigma(\mathscr{A})$ has the cardinality of continuum, while $\mathscr{G}$ has the cardinality of $\mathscr{P}(\mathbb{R})$, since it contains all subsets of Cantor's middle third set (see Exercise 1.8). An example of a non-additive set will be built in Remark 1.23, so that also the second inclusion in (1.11) is strict.

1.6. The Lebesgue measure in $\mathbb{R}$

In this section we build the Lebesgue measure on the real line $\mathbb{R}$. To this aim, we consider first the set $\mathscr{I}$ of all bounded right open intervals of $\mathbb{R}$

$$\mathscr{I} := \{(a, b] \ : \ a, b \in \mathbb{R}, \ a < b\}$$

and the collection $\mathscr{A}$ containing $\varnothing$ and the finite unions of elements of $\mathscr{I}$. Our choice of half-open intervals ensures that $\mathscr{A}$ is a ring, because $\mathscr{I}$ is stable under intersection and relative complement (the families of open and closed intervals, instead, do not have this property).

We define

$$\text{length}((a, b]) := b - a.$$

More generally, any non-empty $A \in \mathscr{A}$ can be written, possibly in many ways, as a disjoint finite union of intervals $I_i, i = 1, \ldots, N$; we define

$$\lambda(A) := \sum_{i=1}^{N} \text{length}(I_i). \tag{1.12}$$

Setting $\lambda(\varnothing) = 0$, it is not hard to show by elementary methods that λ is well defined (*i.e.* $\lambda(A)$ does not depend on the chosen decomposition) and additive on $\mathscr{A}$.

In the next definition we introduce the notion of characteristic function, which can be used to turn set-theoretic operations into algebraic ones: for instance the intersection corresponds to the product, when seen at the level of characteristic functions (see also Exercise 1.1).

Definition 1.19 (Characteristic function of a set). Let $A \subset X$. The characteristic function $\mathbb{1}_A : X \to \{0, 1\}$ is defined by

$$\mathbb{1}_A(x) := \begin{cases} 1 & \text{if } x \in A; \\ 0 & \text{if } x \in X \setminus A. \end{cases}$$

The reader already acquainted with Riemann's theory of integration can also notice that $\lambda(A)$ is the Riemann integral of the characteristic function $\mathbb{1}_A$ of A, and deduce the additivity property of λ directly by the additivity properties of the Riemann integral. In the next theorem we shall rigorously prove these facts, and more. We first state an auxiliary lemma, a simple consequence of the Bolzano-Weierstrass compactness theorem on the real line.

Lemma 1.20. *Any bounded and closed interval J contained in the union of a sequence $\{A_n\}_{n \in \mathbb{N}}$ of open sets is contained in the union of finitely many of them.*

Proof. Assume with no loss of generality that $I = \mathbb{N}$ and $A_n \subset A_{n+1}$, and assume by contradiction that there exist $x_n \in J \setminus A_n$ for all n; by the Bolzano–Weierstrass theorem there exists a subsequence $(x_{n(k)})$ converging to some $x \in J$. If $\bar{n}$ is such that $x \in A_{\bar{n}}$, for k large enough $x_{n(k)}$ belongs to $A_{\bar{n}}$, because $A_{\bar{n}}$ is open. But this is not possible, as soon as $n(k) \geq \bar{n}$, because $x_{n(k)} \notin A_{n(k)}$ and $A_{n(k)} \supset A_{\bar{n}}$. $\qquad\square$

Theorem 1.21. *The set function λ defined in* (1.12) *is σ–additive on $\mathscr{A}$.*

Proof. (λ is well defined) Given disjoint partitions $I_1, \ldots, I_n$ and $J_1, \ldots, J_m$ of $A \in \mathscr{A}$, we say that $J_1, \ldots, J_m$ is finer than $I_1, \ldots, I_n$ if any interval I_i is the disjoint union of some of the intervals J_j. Obviously, given any two partitions, there exists a third partition finer than both: it suffices to take all intersections of elements of the first partition with elements of the second partition, neglecting the empty intersections. Given these remarks, to show that λ is well defined, it suffices to show that $\sum_i \lambda(I_i) = \sum_j \lambda(J_j)$ if $J_1, \ldots, J_m$ is finer than $I_1, \ldots, I_n$. This statement reduces to the fact that $\lambda(I) = \sum_k \lambda(F_k)$ if $I \in \mathscr{I}$ is the disjoint union of some elements $F_k \in \mathscr{I}$; this last statement can be easily proved, starting from the identity $(a, b] = (a, c] \cup (c, b]$, by induction on the number of the intervals F_k.

(λ is additive) If $F, G \in \mathscr{A}$ and $F \cap G = \varnothing$, any disjoint decompositions of F in intervals $I_1, \ldots, I_n \in \mathscr{I}$ and any disjoint decomposition of G in intervals $J_1, \ldots, J_m \in \mathscr{I}$ provide a decomposition $I_1, \ldots, I_n, J_1, \ldots, J_m$ of $F \cup G$ in intervals belonging to $\mathscr{I}$. Using this decomposition to compute $\lambda(F \cup G)$ the additivity easily follows.

(λ is σ–additive) Let $(F_n) \subset \mathscr{A}$ be a sequence of disjoints sets in $\mathscr{A}$ and assume that

$$F := \bigcup_{n=0}^{\infty} F_n \tag{1.13}$$

also belongs to $\mathscr{A}$.

We prove the additivity property first in the case when $F = (x, y] \in \mathscr{I}$. It is also not restrictive to assume that the series $\sum_n \lambda(F_n)$ is convergent. As any F_n is a finite union of intervals, say N_n, we can find, given any $\varepsilon > 0$, a finite union $F'_n \supset F_n$ of intervals in $\mathscr{I}$ such that $\lambda(F'_n) \leq \lambda(F_n) + \varepsilon/2^n$ and the internal part of F'_n contains F_n (just shift the endpoints of each interval in F_n by a small amount, to obtain a larger interval in $\mathscr{I}$, increasing the length at most by $\varepsilon/(N_n 2^n)$). Let also F''_n be the internal part of F'_n, that still includes F_n, and let $x' \in (x, y]$. Then, since $[x', y] \subset \bigcup_n F''_n$, Lemma 1.20 provides an integer k such that $[x', y] \subset \bigcup_0^k F'_n$. Hence, the additivity of λ in $\mathscr{A}$ gives

$$
y - x' \leq \lambda \left(\bigcup_{n=0}^{k} F'_n \right) \leq \sum_{n=0}^{k} \lambda(F'_n)
$$
$$
\leq \sum_{n=0}^{k} \lambda(F_n) + \frac{\varepsilon}{2^n} \leq 2\varepsilon + \sum_{n=0}^{\infty} \lambda(F_n).
$$

By letting first $\varepsilon \downarrow 0$ and then letting $x' \downarrow x$ we obtain that $\lambda(F) \leq \sum_0^{\infty} \lambda(F_n)$. The opposite inequality simply follows by the monotonicity and the additivity of λ, because the finite unions of the sets F_n are contained in F.

In the general case, let

$$
F = \bigcup_{i=1}^{k} I_i,
$$

where $I_1, \ldots, I_k$ are disjoint sets in $\mathscr{I}$. Then, since for any $i \in \{1, \ldots, k\}$ we have that I_i is the disjoint union of $I_i \cap F_n$, we know by the previous step that

$$
\lambda(I_i) = \sum_{n=0}^{\infty} \lambda(I_i \cap F_n).
$$

Adding these identities for $i = 1, \ldots, k$, commuting the sums on the right hand side and eventually using the additivity of λ on $\mathscr{A}$ we obtain

$$
\lambda(F) = \sum_{i=1}^{k} \lambda(I_i \cap F) = \sum_{n=0}^{\infty} \sum_{i=1}^{k} \lambda(I_i \cap F_n) = \sum_{n=0}^{\infty} \lambda(F_n). \qquad \square
$$

We say that a measure μ in $(\mathbb{R}, \mathscr{B}(\mathbb{R}))$ is *translation invariant* if $\mu(A + h) = \mu(A)$ for all $A \in \mathscr{B}(\mathbb{R})$ and $h \in \mathbb{R}$ (notice that, by Exercise 1.2, the class of Borel sets is translation invariant as well). We say also that μ is locally finite if $\mu(I) < \infty$ for all bounded intervals $I \subset \mathbb{R}$.

Theorem 1.22 (Lebesgue measure in $\mathbb{R}$). *There exists a unique, up to multiplication with constants, translation invariant and locally finite measure λ in $(\mathbb{R}, \mathscr{B}(\mathbb{R}))$. The unique such measure λ satisfying $\lambda([0, 1]) = 1$ is called Lebesgue measure.*

Proof. (Existence) Let $\mathscr{A}$ be the class of finite unions of intervals and let $\lambda : \mathscr{A} \to [0, +\infty)$ be the σ–additive set function defined in (1.12). According to Theorem 1.21 λ admits a unique extension, that we still denote by λ, to $\sigma(\mathscr{A}) = \mathscr{B}(\mathbb{R})$. Clearly λ is locally finite, and we can use the uniqueness of the extension to prove translation invariance: indeed, for any $h \in \mathbb{R}$ also the σ–additive measure $A \mapsto \lambda(A+h)$ is an extension of $\lambda|_{\mathscr{A}}$. As a consequence $\lambda(A) = \lambda(A + h)$ for all $h \in \mathbb{R}$.

(Uniqueness) Let ν be a translation invariant and locally finite measure in $(\mathbb{R}, \mathscr{B}(\mathbb{R}))$ and set $c := \nu([0, 1])$. Notice first that the set of atoms of ν is at most countable (Exercise 1.5), and since $\mathbb{R}$ is uncountable there exists at least one x such that $\nu(\{x\}) = 0$. By translation invariance this holds for all x, *i.e.*, ν has no atom.

Excluding the trivial case $c = 0$ (that gives $\nu \equiv 0$ by translation invariance and σ–additivity), we are going to show that $\nu = c\lambda$ on the class $\mathscr{A}$ of finite unions of intervals; by the uniqueness of the extension in Carathéodory theorem this would imply that $\nu = c\lambda$ on $\mathscr{B}(\mathbb{R})$.

By finite additivity and translation invariance it suffices to show that $\nu([0, t)) = ct$ for any $t \geq 0$ (by the absence of atoms the same holds for the intervals $(0, t)$, $(0, t]$, $[0, t]$). Notice first that, for any integer $q \geq 1$, $[0, 1)$ is the union of q disjoint intervals all congruent to $[0, 1/q)$; as a consequence, additivity and translation invariance give

$$\nu\big([0, 1/q)\big) = \frac{\nu([0, 1))}{q} = \frac{c}{q}.$$

Similarly, for any integer $p \geq 1$ the interval $[0, p/q)$ is the union of p disjoint intervals all congruent to $[0, 1/q)$; again additivity and translation invariance give

$$\nu([0, \frac{p}{q})) = p\nu\big([0, \frac{1}{q})\big) = c\frac{p}{q}.$$

By approximation we eventually obtain that $\nu([0, t)) = ct$ for all $t \geq 0$. $\qquad\square$

The completion of the Borel σ–algebra with respect to λ is the so-called σ-algebra of Lebesgue measurable sets. It coincides with the class $\mathscr{C}$ of additive sets with respect to λ^* considered in the proof of Carathéodory theorem (see Exercise 1.12).

Remark 1.23 (Outer Lebesgue measure and non-measurable sets).
The measure λ^* used in the proof of Carathéodory's theorem is also called *outer* Lebesgue measure, and it is defined on all parts of $\mathbb{R}$. The terminology is slightly misleading here, since λ^*, though σ–subadditive, *fails* to be σ–additive. In particular, there exist subsets of $\mathbb{R}$ that are not Lebesgue measurable. To see this, let us consider the equivalence relation in $\mathbb{R}$ defined by $x \sim y$ if $x - y \in \mathbb{Q}$ and let us pick a single element $x \in [0, 1]$ in any equivalence class induced by this relation, thus forming a set $A \subset [0, 1]$. Were this set Lebesgue measurable, all the sets $A + h$

would still be measurable, by translation invariance, and the family of sets $\{A + h\}_{h \in \mathbb{Q}}$ would be a countable and measurable partition of $\mathbb{R}$, with $\lambda^*(A + h) = c$ independent of $h \in \mathbb{Q}$. Now, if $c = 0$ we reach a contradiction with the fact that $\lambda^*(\mathbb{R}) = \infty$, while if $c > 0$ we consider all sets $A + h$ with $h \in \mathbb{Q} \cap [0, 1]$ to obtain

$$2 = \lambda^*([0, 2]) \geq \sum_{h \in \mathbb{Q} \cap [0,1]} \lambda^*(A + h) = \infty,$$

reaching again a contradiction.

Notice that this example is not constructive and strongly requires the axiom of choice (also the arguments based on cardinality, see Exercise 1.19 and Exercise 1.20, have this limitation). On the other hand, one can give *constructive* examples of Lebesgue measurable sets that are not Borel (see for instance 2.2.11 in [3]).

The construction done in the previous remark rules out the existence of locally finite and translation invariant σ–additive measures defined on *all* parts of $\mathbb{R}$. In $\mathbb{R}^n$, with $n \geq 3$, the famous *Banach–Tarski paradox* (see for instance [6]) shows that it is also impossible to have a locally finite, invariant under rigid motions and *finitely* additive measure defined on all parts of $\mathbb{R}^n$.

1.7. Inner and outer regularity of measures on metric spaces

Let (E, d) be a metric space and let μ be a finite measure on $(E, \mathscr{B}(E))$. We shall prove a regularity property of μ.

Proposition 1.24. *For any $B \in \mathscr{B}(E)$ we have*

$$\mu(B) = \sup\{\mu(C) : \ C \subset B, \ \text{closed}\} = \inf\{\mu(A) : \ A \supset B, \ \text{open}\}.$$
$$(1.14)$$

Proof. Let us set

$$\mathscr{K} = \{B \in \mathscr{B}(E) : \ (1.14) \ \text{holds}\}.$$

It is enough to show that $\mathscr{K}$ is a σ–algebra of parts of E including the open sets of E. Obviously $\mathscr{K}$ contains E and $\varnothing$. Moreover, if $B \in \mathscr{K}$ then its complement B^c belongs to $\mathscr{K}$. Let us prove now that $(B_n) \subset \mathscr{K}$ implies $\bigcup_n B_n \in \mathscr{K}$. Fix $\varepsilon > 0$. We are going to show that there exist a closed set C and an open set A such that

$$C \subset \bigcup_{n=0}^{\infty} B_n \subset A, \qquad \mu(A \setminus C) \leq 2\varepsilon. \qquad (1.15)$$

Let $n \in \mathbb{N}$. Since $B_n \in \mathcal{K}$ there exist an open set A_n and a closed set C_n such that $C_n \subset B_n \subset A_n$ and

$$\mu(A_n \setminus C_n) \leq \frac{\varepsilon}{2^{n+1}}.$$

Setting $A := \bigcup_n A_n$, $S := \bigcup_n C_n$ we have $S \subset \bigcup_n B_n \subset A$ and $\mu(A \setminus S) \leq \varepsilon$. However, A is open but S is not necessarily closed. So, we approximate S by setting $S_n := \bigcup_0^n C_k$. The set S_n is obviously closed, $S_n \uparrow S$ and consequently $\mu(S_n) \uparrow \mu(S)$. Therefore there exists $n_\varepsilon \in \mathbb{N}$ such that $\mu(S \setminus S_{n_\varepsilon}) < \varepsilon$. Now, setting $C = S_{n_\varepsilon}$ we have $C \subset \bigcup_n B_n \subset A$ and $\mu(A \setminus C) < \mu(A \setminus S) + \mu(S \setminus C) < 2\varepsilon$. Therefore $\bigcup_n B_n \in \mathcal{K}$. We have proved that $\mathcal{K}$ is a σ–algebra. It remains to show that $\mathcal{K}$ contains the open subsets of E. In fact, let A be open and set

$$C_n = \left\{ x \in E : \ d(x, A^c) \geq \frac{1}{n} \right\},$$

where $d(x, A^c) := \inf_{y \in A^c} d(x, y)$ is the distance function from A^c. Then C_n are closed subsets of A, and moreover $C_n \uparrow A$, which implies $\mu(A \setminus C_n) \downarrow 0$. Thus the conclusion follows. $\qquad\square$

Notice that inner and outer approximation hold for μ–measurable sets B as well: one has just to notice that there exist Borel sets B_1, B_2 such that $B_1 \subset B \subset B_2$ with $\mu(B_2 \setminus B_1) = 0$, and apply inner approximation to B_1 and outer approximation to B_2.

Remark 1.25 (Inner and outer approximation for σ-finite measures). It is possible to extend the inner approximation property to σ-finite measures: suffices to assume the existence of a sequence of closed sets C_n with finite measure such that $\mu(X \setminus \cup_n C_n) = 0$. Indeed, assuming with no loss of generality that $C_n \subset C_{n+1}$, we know that for any Borel set B and any $n \in \mathbb{N}$ it holds

$$\mu(B \cap C_n) = \sup \{\mu(C) : \ C \text{ closed}, C \subset B \cap C_n\},$$

so that

$$\mu(B \cap C_n) \leq \sup \{\mu(C) : \ C \text{ closed}, C \subset B\}.$$

Letting $n \uparrow \infty$ we recover the inner approximation property. Analogously, if we assume the existence of a sequence of open sets A_n with finite measure satisfying $X = \cup_n A_n$, we have the outer approximation property: indeed, for any n and any $\epsilon > 0$ we can find (assuming with no loss of generality $\mu(B) < +\infty$) open sets $B_n \subset A_n$ containing

$B \cap A_n$ and such that $\mu\big(B_n \setminus (B \cap A_n)\big) < \epsilon 2^{-n}$. It follows that $\cup_n B_n$ contains B and

$$\mu\left(\bigcup_{n \in \mathbb{N}} B_n \setminus B\right) < 2\epsilon.$$

Since B_n are also open in X, the set $\cup_n B_n$ is open and since ϵ is arbitrary we get the outer approximation property.

We conclude this chapter with the following result, whose proof is a straightforward consequence of Proposition 1.24 (alternatively, one can use Dynkin's argument, since the class of closed sets is a π-system and generates the Borel σ-algebra).

Corollary 1.26. *Let* μ, ν *be finite measures in* $(E, \mathscr{B}(E))$, *such that* $\mu(C) = \nu(C)$ *for any closed subset C of E. Then $\mu = \nu$.*

Exercises

1.1 Given $A \subset X$, denote by $\mathbb{1}_A : X \to \{0, 1\}$ its characteristic function, equal to 1 on A and equal to 0 on A^c. Show that

$$\mathbb{1}_{A \cup B} = \max\{\mathbb{1}_A, \mathbb{1}_B\}, \quad \mathbb{1}_{A \cap B} = \min\{\mathbb{1}_A, \mathbb{1}_B\}, \quad \mathbb{1}_{A^c} = \mathbb{1}_X - \mathbb{1}_A$$

and that

$$\limsup_{n \to \infty} A_n = A \quad \Longleftrightarrow \quad \limsup_{n \to \infty} \mathbb{1}_{A_n} = \mathbb{1}_A,$$

$$\liminf_{n \to \infty} A_n = A \quad \Longleftrightarrow \quad \liminf_{n \to \infty} \mathbb{1}_{A_n} = \mathbb{1}_A.$$

1.2 Let $A \subset \mathbb{R}^n$ be a Borel set. Show that for $h \in \mathbb{R}^n$ and $t \in \mathbb{R}$ the sets

$$A + h := \{a + h \,:\, a \in A\}, \qquad tA := \{ta \,:\, a \in A\}$$

are Borel as well.

1.3 Find an example of a σ–additive measure μ on a σ–algebra $\mathscr{A}$ such that there exist $A_n \in \mathscr{A}$ with $A_n \downarrow A$ and $\inf_n \mu(A_n) > \mu(A)$.

1.4 Let μ be additive and finite, on an algebra $\mathscr{A}$. Show that μ is σ–additive if and only if it is continuous along nonincreasing sequences.

1.5 Let μ be a finite measure on $(X, \mathscr{E})$. Show that the set of atoms of μ, defined by

$$A_\mu := \{x \in X \,:\, \{x\} \in \mathscr{E} \text{ and } \mu(\{x\}) > 0\}$$

is at most countable. Show that the same is true for σ–finite measures, and provide an example of a measure space for which this property fails.

1.6 Let $(X, \mathscr{E}, \mu)$ be a measure space, with μ finite. We say that μ is diffuse if for all $A \in \mathscr{E}$ with $\mu(A) > 0$ there exists $B \subset A$ with $0 < \mu(B) < \mu(A)$. Show that, if μ is diffuse, then $\mu(\mathscr{E}) = [0, \mu(X)]$.

1.7 Show that if X is a separable metric space and $\mathscr{E}$ is the Borel σ–algebra, then a σ–additive measure $\mu : \mathscr{E} \to [0, +\infty)$ is diffuse if and only if μ has no atom.

1.8 Let λ be the Lebesgue measure in $[0, 1]$. Show the existence of a λ–negligible set having the cardinality of the continuum. *Hint:* consider the classical Cantor's middle third set, obtained by removing the interval $(1/3, 2/3)$ from $[0, 1]$, then by removing the intervals $(1/9, 2/9)$ and $(7/9, 8/9)$, and so on.

1.9 Let λ be the Lebesgue measure in $[0, 1]$. Show the existence, for any $\varepsilon > 0$, of a closed set $C \subset [0, 1]$ containing no interval and such that $\lambda(C) > 1 - \varepsilon$. *Hint:* remove from $[0, 1]$ a sequence of open intervals, centered on the rational points of $[0, 1]$.

1.10 Using the previous exercise, write $[0, 1] = A \cup B$ where A is negligible in the measure-theoretic sense (*i.e.* $\lambda(A) = 0$) and B is negligible in the Baire category sense (*i.e.* it is the union of countably many closed sets with empty interior). So, the two concepts of negligible should be never used at the same time.

1.11 $\star$ Let λ be the Lebesgue measure in $[0, 1]$. Construct a Borel set $E \subset (0, 1)$ such that

$$0 < \lambda(E \cap I) < \lambda(I)$$

for any open interval $I \subset (0, 1)$.

1.12 Let $(X, \mathscr{E}, \mu)$ be a measure space and let $\mu^* : \mathscr{P}(X) \to [0, +\infty]$ be the outer measure induced by μ. Show that the completed σ–algebra $\mathscr{E}_\mu$ is contained in the class $\mathscr{C}$ of additive sets with respect to μ^*.

1.13 Let $(X, \mathscr{E}, \mu)$ be a measure space and let $\mu^* : \mathscr{P}(X) \to [0, +\infty]$ be the outer measure induced by μ. Show that for all $A \subset X$ there exists $B \in \mathscr{E}$ containing A with $\mu(B) = \mu^*(A)$.

1.14 Let $(X, \mathscr{E}, \mu)$ be a measure space. Check the following statements, made in Definition 1.12:

(i) $\mathscr{E}_\mu$ is a σ–algebra;
(ii) the extension $\mu(A) := \mu(B)$, where $B \in \mathscr{E}$ is any set such that $A \triangle B$ is contained in a μ–negligible set of $\mathscr{E}$, is well defined and σ–additive on $\mathscr{E}_\mu$;
(iii) μ–negligible sets of $\mathscr{E}_\mu$ are characterized by the property of being contained in a μ–negligible set of $\mathscr{E}$.

1.15 $\star$ Let $(X, \mathscr{E}, \mu)$ be a measure space and let $\mu^* : \mathscr{P}(X) \to [0, +\infty]$ be the outer measure induced by μ. Show that if $\mu(X)$ is finite, the class $\mathscr{C}$ of additive sets with respect to μ^* coincides with the class of $\mathscr{E}_\mu$–measurable sets. *Hint:* one inclusion is provided by Exercise 1.12. For the other one, given an additive set A, by applying Exercise 1.13 twice, find first a set $B \in \mathscr{E}$ with $\mu^*(B \setminus A) = 0$, and then a set $C \in \mathscr{E}$ with $\mu(C) = 0$ and $B \setminus A \subset C$.

1.16 Find a σ–algebra $\mathscr{E} \subset \mathscr{P}(\mathbb{N})$ containing infinitely many sets and such that any $B \in \mathscr{E}$ different from $\varnothing$ has an infinite cardinality.

1.17 Find $\mu : \mathscr{P}(\mathbb{N}) \to \{0, +\infty\}$ that is additive, but not σ–additive.

1.18 $\star$ Let ω be the first uncountable ordinal and, for $\mathscr{K} \subset \mathscr{P}(X)$, define by transfinite induction a family $\mathscr{F}^{(i)}$, $i \in \omega$, as follows: $\mathscr{F}^{(0)} := \mathscr{K} \cup \{\varnothing\}$,

$$\mathscr{F}^{(i)} := \left\{ \bigcup_{k=0}^{\infty} A_k, \ B^c \ : \ (A_k) \subset \mathscr{F}^{(j)}, \ B \in \mathscr{F}^{(j)} \right\},$$

if i is the successor of j, and $\mathscr{F}^{(i)} := \bigcup_{j \in i} \mathscr{F}^{(j)}$ otherwise.
Show that $\bigcup_{i \in \omega} \mathscr{F}^{(i)} = \sigma(\mathscr{K})$.

1.19 $\star$ Show that $\mathscr{B}(\mathbb{R})$ has the cardinality of the continuum. *Hint*: use the construction of the previous exercise, and the fact that ω has at most the cardinality of continuum.

1.20 $\star$ Show that the σ–algebra $\mathscr{L}$ of Lebesgue measurable sets has the same cardinality of $\mathscr{P}(\mathbb{R})$, thus strictly greater than the continuum. *Hint*: consider all subsets of Cantor's middle third set.

1.21 $\star\star$ Show that the cardinality of any σ–algebra is either finite or uncountable.

1.22 Let X be a set and let $\mathscr{A} \subset \mathscr{P}(X)$ be an algebra with finite cardinality. Show that its cardinality is equal to 2^n for some integer $n \geq 1$.

1.23 $\star$ Let $(X, \mathscr{E}, \mu)$ be a a measure space and suppose that X is finite or countable. Show the existence of a measure $\tilde{\mu}$ on $\mathscr{P}(X)$ that extends μ, that is, $\mu(A) = \tilde{\mu}(A)$ for all $A \in \mathscr{E}$.

1.24 $\star\star$ Find an example of an additive set function $\mu : \mathscr{P}(\mathbb{N}) \to \{0, 1\}$, with $\mu(\mathbb{N}) = 1$ and $\mu(\{n\}) = 0$ for all $n \in \mathbb{N}$ (in particular μ is not σ–additive, the construction of this example requires Zorn's lemma).

1.25 $\star$ Let $C \in \mathscr{B}([0, 1])$ with $\lambda(C) > 0$. Without using the continuum hypothesis, show that C has the cardinality of continuum.

1.26 $\star$ Let (K, d) be a compact metric space and let μ be as in Exercise 1.24. Let's say that a sequence $(x_n) \subset K$ μ-converges to $x \in K$ if

$$\mu\big(\{n \in \mathbb{N} : d(x_n, x) > \varepsilon\}\big) = 0 \qquad \forall \varepsilon > 0.$$

Show that any sequence $(x_n) \subset K$ is μ-convergent and that the μ-limit is unique.

Chapter 2
Integration

This chapter is devoted to the construction of the integral of $\mathscr{E}$–measurable functions in general measure spaces $(\Omega, \mathscr{E}, \mu)$, and its main continuity and lower semicontinuity properties. Having built in the previous chapter the Lebesgue measure in the real line $\mathbb{R}$, we obtain as a byproduct the Lebesgue integral on $\mathbb{R}$; in the last section we compare Lebesgue and Riemann integral.

In the construction of the integral we prefer to empahsize two viewpoints: the first, more traditional one

$$\int_X f \, d\mu = \sum_{z \in \mathrm{Im}(f)} z \mu(\{f = z\})$$

is appropriate to deal with simple functions (*i.e.* functions whose range is finite) and useful to show the additivity of the integral with respect to f. The second one, for nonnegative functions is summarized by the formula

$$\int_X f \, d\mu = \int_0^\infty \mu(\{f > t\}) \, dt.$$

This second viewpoint is more appropriate to show the continuity properties of the integral with respect to f (the integral on the right side can be elementarily defined, since $t \mapsto \mu(\{f > t\})$ is nonincreasing, see Section 2.4.3). Of course we show that the two viewpoints are consistent if we restrict ourselves to the class of simple functions.

2.1. Inverse image of a function

Let X be a non empty set. For any function $\varphi \colon X \to Y$ and any $I \in \mathscr{P}(Y)$ we set

$$\varphi^{-1}(I) := \{x \in X : \varphi(x) \in I\} = \{\varphi \in I\}.$$

The set $\varphi^{-1}(I)$ is called the *inverse image* of I.

Let us recall some elementary properties of φ^{-1} (the easy proofs are left to the reader as an exercise):

(i) $\varphi^{-1}(I^c) = (\varphi^{-1}(I))^c$ for all $I \in \mathscr{P}(Y)$;

(ii) if $\{J_i\}_{i \in I} \subset \mathscr{P}(Y)$ we have

$$\bigcup_{i \in I} \varphi^{-1}(J_i) = \varphi^{-1}\left(\bigcup_{i \in I} J_i\right), \qquad \bigcap_{i \in I} \varphi^{-1}(J_i) = \varphi^{-1}\left(\bigcap_{i \in I} J_i\right).$$

In particular, if $I \cap J = \varnothing$ we have $\varphi^{-1}(I) \cap \varphi^{-1}(J) = \varnothing$. Also, if $\mathscr{E} \subset \mathscr{P}(Y)$ and we consider the family $\varphi^{-1}(\mathscr{E})$ of subset of X defined by

$$\varphi^{-1}(\mathscr{E}) := \left\{\varphi^{-1}(I) : \ I \in \mathscr{E}\right\}, \tag{2.1}$$

we have that $\varphi^{-1}(\mathscr{E})$ is a σ–algebra whenever $\mathscr{E}$ is a σ–algebra.

2.2. Measurable and Borel functions

We are given measurable spaces $(X, \mathscr{E})$ and $(Y, \mathscr{F})$. We say that a function $\varphi \colon X \to Y$ is $(\mathscr{E}, \mathscr{F})$–*measurable* if $\varphi^{-1}(\mathscr{F}) \subset \mathscr{E}$. If $(Y, \mathscr{F}) = (\mathbb{R}, \mathscr{B}(\mathbb{R}))$, we say that φ is a real valued $\mathscr{E}$–*measurable* function, and if (X, d) is a metric space and $\mathscr{E}$ is the Borel σ–algebra, we say that φ is a real valued *Borel* function.

The following simple but useful proposition shows that the measurability condition needs to be checked only on a class of generators.

Proposition 2.1 (Measurability criterion). *Let* $\mathscr{G} \subset \mathscr{F}$ *be such that* $\sigma(\mathscr{G}) = \mathscr{F}$. *Then* $\varphi \colon X \to Y$ *is* $(\mathscr{E}, \mathscr{F})$–*measurable if and only if* $\varphi^{-1}(I) \in \mathscr{E}$ *for all* $I \in \mathscr{G}$ *(equivalently, iff* $\varphi^{-1}(\mathscr{G}) \subset \mathscr{E}$*).*

Proof. Consider the family $\mathscr{D} := \{I \in \mathscr{F} : \varphi^{-1}(I) \in \mathscr{E}\}$. By the above-mentioned properties of φ^{-1} as an operator between $\mathscr{P}(Y)$ and $\mathscr{P}(X)$, it follows that $\mathscr{D}$ is a σ–algebra including $\mathscr{G}$. So, it coincides with $\sigma(\mathscr{G}) = \mathscr{F}$. $\qquad\square$

A simple consequence of the previous proposition is the fact that any continuous function is a Borel function: more precisely, assume that $\varphi : X \to Y$ is continuous and that $\mathscr{E} = \mathscr{B}(X)$ and $\mathscr{F} = \mathscr{B}(Y)$. Then, the σ–algebra

$$\left\{A \subset Y : \varphi^{-1}(A) \in \mathscr{B}(X)\right\}$$

contains the open subsets of Y (as, by the continuity of φ, $\varphi^{-1}(A)$ is open in X, and in particular Borel, whenever A is open in Y), and then it contains the generated σ–algebra, *i.e.* $\mathscr{B}(Y)$.

The following proposition, whose proof is straightforward, shows that the class of measurable functions is stable under composition.

Proposition 2.2. *Let $(X, \mathscr{E})$, $(Y, \mathscr{F})$, $(Z, \mathscr{G})$ be measurable spaces and let $\varphi : X \to Y$ and $\psi : Y \to Z$ be respectively $(\mathscr{E}, \mathscr{F})$–measurable and $(\mathscr{F}, \mathscr{G})$–measurable. Then $\psi \circ \varphi$ is $(\mathscr{E}, \mathscr{G})$–measurable.*

It is often convenient to consider functions with values in the extended space $\overline{\mathbb{R}} := \mathbb{R} \cup \{+\infty, -\infty\}$, the so-called *extended functions*. We say that a mapping $\varphi \colon X \to \overline{\mathbb{R}}$ is $\mathscr{E}$–measurable if

$$\varphi^{-1}(\{-\infty\}),\ \varphi^{-1}(\{+\infty\}) \in \mathscr{E} \qquad \text{and} \qquad \varphi^{-1}(I) \in \mathscr{E},\ \forall I \in \mathscr{B}(\mathbb{R}). \tag{2.2}$$

This condition can also be interpreted in terms of measurability between $\mathscr{E}$ and a suitable Borel σ–algebra in $\overline{\mathbb{R}}$, see Exercise 2.3. Analogously, when (X, d) is a metric space and $\mathscr{E}$ is the Borel σ–algebra, we say that $\varphi \colon X \to \overline{\mathbb{R}}$ is Borel whenever the conditions above hold.

The following proposition shows that extended $\mathscr{E}$–measurable functions are stable under pointwise limits and countable supremum and infimum.

Proposition 2.3. *Let (φ_n) be a sequence of extended $\mathscr{E}$–measurable functions. Then the following functions are $\mathscr{E}$–measurable:*

$$\sup_{n \in \mathbb{N}} \varphi_n(x), \quad \inf_{n \in \mathbb{N}} \varphi_n(x), \quad \limsup_{n \to \infty} \varphi_n(x), \quad \liminf_{n \to \infty} \varphi_n(x).$$

Proof. Let us prove that $\varphi(x) := \sup_n \varphi_n(x)$ is $\mathscr{E}$–measurable (all other cases can be deduced from this one, or directly proved by similar arguments). For any $a \in \overline{\mathbb{R}}$ we have

$$\varphi^{-1}([-\infty, a]) = \bigcap_{n=0}^{\infty} \varphi_n^{-1}([-\infty, a]) \in \mathscr{E}.$$

In particular $\{\varphi = -\infty\} \in \mathscr{E}$, so that $\varphi^{-1}((-\infty, a]) \in \mathscr{E}$ for all $a \in \mathbb{R}$; by letting $a \uparrow \infty$ we get $\varphi^{-1}(\mathbb{R}) \in \mathscr{E}$. As a consequence, the class

$$\left\{ I \in \mathscr{B}(\mathbb{R}) \ : \ \varphi^{-1}(I) \in \mathscr{E} \right\}$$

is a σ–algebra containing the intervals of the form $(-\infty, a]$ with $a \in \mathbb{R}$, and therefore coincides with $\mathscr{B}(\mathbb{R})$. Eventually, $\{\varphi = +\infty\} = X \setminus [\varphi^{-1}(\mathbb{R}) \cup \{\varphi = -\infty\}]$ belongs to $\mathscr{E}$ as well. $\qquad \square$

2.3. Partitions and simple functions

Let $(X, \mathscr{E})$ be a measurable space. A function $\varphi \colon X \to \mathbb{R}$ is said to be *simple* if its range $\varphi(X)$ is a finite set. The class of simple functions is obviously a real vector space, as the range of $\varphi + \psi$ is contained in

$$\{a + b \ : \ a \in \operatorname{range}(\varphi),\ b \in \operatorname{range}(\psi)\}.$$

If $\varphi(X) = \{a_1, \ldots, a_n\}$, with $a_i \neq a_j$ if $i \neq j$, setting $A_i = \varphi^{-1}(\{a_i\})$, $i = 1, \ldots, n$ we can canonically represent φ as

$$\varphi(x) = \sum_{k=1}^{n} a_k \mathbb{1}_{A_k}, \quad x \in X. \tag{2.3}$$

Moreover, $A_1, \ldots, A_n$ is a *finite partition* of X (*i.e.* A_i are mutually disjoint and their union is equal to X). However, a simple function φ has many representations of the form

$$\varphi(x) = \sum_{k=1}^{N} a_k' \mathbb{1}_{A_k'}, \quad x \in X,$$

where $A_1', \ldots, A_N'$ need not be mutually disjoint and a_k' need not be in the range of φ. For instance

$$\mathbb{1}_{[0,1)} + 3\mathbb{1}_{[1,2]} = \mathbb{1}_{[0,2]} + 2\mathbb{1}_{[1,2]}.$$

It is easy to check that a simple function is $\mathcal{E}$–measurable if, and only if, all level sets A_k in (2.3) are $\mathcal{E}$–measurable; in this case we shall also say that $\{A_k\}$ is a *finite $\mathcal{E}$–measurable partition* of X.

Now we show that any nonnegative $\mathcal{E}$–measurable function can be approximated by simple functions; a variant of this result, with a different construction, is proposed in Exercise 2.8.

Proposition 2.4. *Let φ be a nonnegative extended $\mathcal{E}$–measurable function. For any $n \in \mathbb{N}^*$, define*

$$\varphi_n(x) = \begin{cases} \dfrac{i-1}{2^n} & \text{if } \dfrac{i-1}{2^n} \leq \varphi(x) < \dfrac{i}{2^n}, \ i = 1, 2, \ldots, n2^n; \\[2mm] n & \text{if } \varphi(x) \geq n. \end{cases} \tag{2.4}$$

Then φ_n are simple and $\mathcal{E}$–measurable, (φ_n) is nondecreasing and convergent to φ. If in addition φ is bounded the convergence is uniform.

Proof. It is not difficult to check that (φ_n) is nondecreasing. Moreover, we have

$$0 \leq \varphi(x) - \varphi_n(x) \leq \frac{1}{2^n} \quad \text{if } \varphi(x) < n, \ x \in X,$$

and

$$0 \leq \varphi(x) - \varphi_n(x) = \varphi(x) - n \quad \text{if } \varphi(x) \geq n, \ x \in X.$$

So, the conclusion easily follows. $\qquad\square$

2.4. Integral of a nonnegative $\mathscr{E}$–measurable function

We are given a measure space $(X, \mathscr{E}, \mu)$. We start to define the integral for simple nonnegative functions.

2.4.1. Integral of simple functions

Let φ be a nonnegative simple $\mathscr{E}$–measurable function, and let us represent it as

$$\varphi(x) = \sum_{k=1}^{N} a_k \mathbb{1}_{A_k}, \quad x \in X,$$

with $N \in \mathbb{N}$, $a_1, \ldots, a_N \geq 0$ and $A_1, \ldots, A_N$ in $\mathscr{E}$. Then we define (using the standard convention in the theory of integration that $0 \cdot \infty = 0$),

$$\int_X \varphi \, d\mu := \sum_{k=1}^{N} a_k \mu(A_k).$$

It is easy to see that the definition does not depend on the choice of the representation formula for φ. Indeed, let $\{b_1, \ldots, b_M\}$ be the range of φ and let $\varphi = \sum_1^M b_j \mathbb{1}_{B_j}$, with $B_j := \varphi^{-1}(b_j)$, be the canonical representation of φ. We have to prove that

$$\sum_{k=1}^{N} a_k \mu(A_k) = \sum_{j=1}^{M} b_j \mu(B_j). \tag{2.5}$$

As the B_i's are pairwise disjoint, (2.5) follows by adding the M identities

$$\sum_{k=1}^{N} a_k \mu(A_k \cap B_j) = b_j \mu(B_j) \qquad j = 1, \ldots, M. \tag{2.6}$$

In order to show (2.6) we fix j and consider, for $I \subset \{1, \ldots, N\}$, the sets

$$A_I := \big\{ x \in B_j \ : \ x \in A_i \text{ iff } i \in I \big\},$$

so that $\{A_I\}$ are a $\mathscr{E}$–measurable partition of B_j and $x \in A_I$ iff the set of i's for which $x \in A_i$ coincides with I. Then, using first the fact that $A_I \subset A_i$ if $i \in I$, and $A_i \cap A_I = \varnothing$ otherwise, and then the fact that $\sum_{k \in I} a_k = b_j$ whenever $A_I \neq \varnothing$ (because $\sum_1^N a_k \mathbb{1}_{A_k}$ coincides with b_j, the constant value of φ on B_j), we have

$$\sum_{k=1}^{N} a_k \mu(A_k \cap B_j) = \sum_{k=1}^{N} \sum_I a_k \mu(A_k \cap A_I) = \sum_I \sum_{k=1}^{N} a_k \mu(A_k \cap A_I)$$
$$= \sum_I \sum_{k \in I} a_k \mu(A_I) = \sum_I b_j \mu(A_I) = b_j \mu(B_j).$$

Proposition 2.5. *Let φ, ψ be simple nonnegative $\mathscr{E}$–measurable functions on X and let α, $\beta \geq 0$. Then $\alpha\varphi + \beta\psi$ is simple, $\mathscr{E}$–measurable and we have*

$$\int_X (\alpha\varphi + \beta\psi)\, d\mu = \alpha \int_X \varphi\, d\mu + \beta \int_X \psi\, d\mu$$

Proof. Let

$$\varphi = \sum_{k=1}^n a_k \mathbb{1}_{A_k}, \qquad \psi = \sum_{h=1}^m b_h \mathbb{1}_{B_h}$$

with $\{A_k\}$, $\{B_h\}$ finite $\mathscr{E}$–measurable partitions of X. Then $\{A_k \cap B_h\}$ is a finite $\mathscr{E}$–measurable partition of X and $\alpha\varphi + \beta\psi$ is constant (and equal to $\alpha a_k + \beta b_h$) on any element $A_k \cap B_h$ of the partition. Therefore the level sets of $\alpha\varphi + \beta\psi$ are finite unions of elements of this partition and the $\mathscr{E}$–measurability of $\alpha\varphi + \beta\psi$ follows (see also Exercise 2.2). Then, writing

$$\varphi(x) = \sum_{k=1}^n \sum_{h=1}^m a_k \mathbb{1}_{A_k \cap B_h}(x), \quad \psi(x) = \sum_{k=1}^n \sum_{h=1}^m b_h \mathbb{1}_{A_k \cap B_h}(x), \quad x \in X,$$

we arrive at the conclusion. $\qquad\qquad\square$

2.4.2. The repartition function

Let $\varphi \colon X \to \overline{\mathbb{R}}$ be $\mathscr{E}$–measurable. The *repartition function F* of φ, relative to μ, is defined by

$$F(t) := \mu(\{\varphi > t\}), \qquad t \in \mathbb{R}.$$

The function F is nonincreasing and satisfies

$$\lim_{t \to -\infty} F(t) = \lim_{n \to -\infty} F(n) = \lim_{n \to \infty} \mu(\{\varphi > -n\}) = \mu(\{\varphi > -\infty\}),$$

and, if μ is finite,

$$\lim_{t \to +\infty} F(t) = \lim_{n \to \infty} F(n) = \lim_{n \to \infty} \mu(\{\varphi > n\}) = \mu(\{\varphi = +\infty\}),$$

since

$$\{\varphi > -\infty\} = \bigcup_{n=1}^\infty \{\varphi > -n\}, \qquad \{\varphi = +\infty\} = \bigcap_{n=1}^\infty \{\varphi > n\}.$$

Other important properties of F are provided by the following result.

Proposition 2.6. *Let* $\varphi\colon X \to \overline{\mathbb{R}}$ *be* $\mathscr{E}$*–measurable and let* F *be its re-partition function.*

(i) *For any* $t_0 \in \mathbb{R}$ *we have* $\lim\limits_{t \to t_0^+} F(t) = F(t_0)$, *that is,* F *is right continuous.*

(ii) *If* μ *is finite, for any* $t_0 \in \mathbb{R}$ *we have* $\lim\limits_{t \to t_0^-} F(t) = \mu(\{\varphi \geq t_0)\}$, *that is,* F *has left limits* [1].

Proof. Let us prove (i). We have

$$\lim_{t \to t_0^+} F(t) = \lim_{n \to +\infty} F\left(t_0 + \frac{1}{n}\right) = \lim_{n \to +\infty} \mu\left(\left\{\varphi > t_0 + \frac{1}{n}\right\}\right)$$
$$= \mu(\{\varphi > t_0\}) = F(t_0),$$

since

$$\{\varphi > t_0\} = \bigcup_{n=1}^{\infty} \left\{\varphi > t_0 + \frac{1}{n}\right\} = \lim_{n \to \infty} \left\{\varphi > t_0 + \frac{1}{n}\right\}.$$

So, (i) follows. We prove now (ii). We have

$$\lim_{t \to t_0^-} F(t) = \lim_{n \to +\infty} F\left(t_0 - \frac{1}{n}\right)$$
$$= \lim_{n \to +\infty} \mu\left(\left\{\varphi > t_0 - \frac{1}{n}\right\}\right) = \mu(\{\varphi \geq t_0\}),$$

since

$$\{\varphi \geq t_0\} = \bigcap_{n=1}^{\infty} \left\{\varphi > t_0 - \frac{1}{n}\right\} = \lim_{n \to \infty} \left\{\varphi > t_0 - \frac{1}{n}\right\}$$

and (ii) follows. $\qquad\qquad\square$

From Proposition 2.6 it follows that, in the case when μ is finite, F is continuous at t_0 iff $\mu(\{\varphi = t_0\}) = 0$.

Now we want to extend the integral operator to nonnegative $\mathscr{E}$–measurable functions. Let φ be a nonnegative, simple and $\mathscr{E}$–measurable function and let

$$\varphi(x) = \sum_{k=0}^{n} a_k \mathbb{1}_{A_k}, \quad x \in X,$$

---

[1] In the literature F is called a *cadlag* function.

with $n \in \mathbb{N}^*$, $0 = a_0 < a_1 < a_2 < \cdots < a_n < \infty$. Then the repartition function F of φ is given by

$$F(t) = \begin{cases} \mu(A_1) + \mu(A_2) + \cdots + \mu(A_n) = F(0) & \text{if } 0 \le t < a_1 \\ \mu(A_2) + \mu(A_3) + \cdots + \mu(A_n) = F(a_1) & \text{if } a_1 \le t < a_2 \\ \cdots\cdots\cdots\cdots\cdots\cdots\cdots\cdots\cdots\cdots\cdots\cdots\cdots\cdots\cdots\cdots\cdots\cdots\cdots \\ \mu(A_n) = F(a_{n-1}) & \text{if } a_{n-1} \le t < a_n \\ 0 = F(a_n) & \text{if } t \ge a_n. \end{cases}$$

Consequently, we can write

$$\begin{aligned}
\int_X \varphi(x)\, d\mu(x) &= \sum_{k=1}^{n} a_k \mu(A_k) = \sum_{k=1}^{n} a_k (F(a_{k-1}) - F(a_k)) \\
&= \sum_{k=1}^{n} a_k F(a_{k-1}) - \sum_{k=1}^{n} a_k F(a_k) \\
&= \sum_{k=0}^{n-1} a_{k+1} F(a_k) - \sum_{k=0}^{n-1} a_k F(a_k) \\
&= \sum_{k=0}^{n-1} (a_{k+1} - a_k) F(a_k) = \int_0^\infty F(t)\, dt.
\end{aligned}$$

(2.7)

Example 2.7. We set $X = \mathbb{R}$, $\mu = \lambda$,

$$A_1 = [1, 2] \cup [10, 11], \quad A_2 = [2, 3], \quad A_3 = [3, 4],$$
$$A_4 = [4, 6], \quad A_5 = [7, 10],$$
$$a_1 = 5, \quad a_2 =, \quad a_3 = 10, \quad a_4 = 7, \quad a_5 = 2$$

and $\varphi := \sum_{k=1}^{5} a_k \mathbb{1}_{A_k}$ to be the simple function shown in Figure 2.1. It is easy to verify that F has the graph shown in the right picture in Figure 2.1.

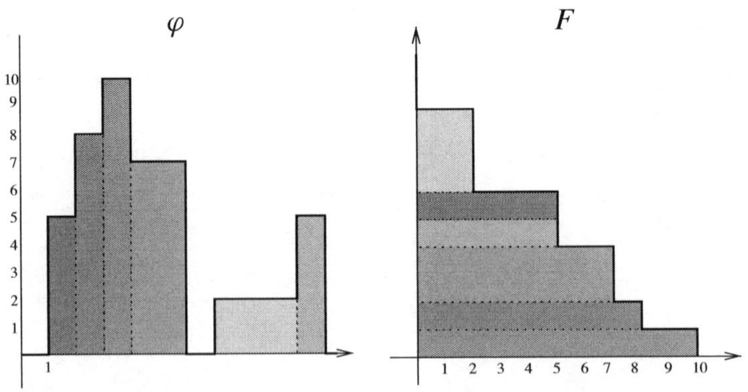

Figure 2.1. a simple function φ, and its repartition F

The color scheme used for the areas below the two graphs in 2.1 proves graphically that the areas are identical.

Now, we want to define the integral of any nonnegative extended $\mathscr{E}-$ measurable function by generalizing formula (2.7). For this, we need first to define the integral of any nonnegative nonincreasing function in $(0, +\infty)$.

2.4.3. The archimedean integral

We generalize here the (inner) Riemann integral to any nonincreasing function $f : [0, +\infty) \to [0, +\infty]$. The strategy is to consider the supremum of the areas of piecewise constant minorants of f.

Let Σ be the set of all finite decompositions $\sigma = \{t_1, \ldots, t_N\}$ of $[0, +\infty]$, where $N \in \mathbb{N}^*$ and $0 = t_0 \le t_1 < \cdots < t_N < +\infty$.

Let now $f : [0, +\infty) \to [0, +\infty]$ be a nonincreasing function. For any $\sigma = \{t_0, t_1, \ldots, t_N\} \in \Sigma$ we consider the partial sum

$$I_f(\sigma) := \sum_{k=0}^{N-1} f(t_{k+1})(t_{k+1} - t_k). \tag{2.8}$$

We define

$$\int_0^\infty f(t)\, dt := \sup\{I_f(\sigma) : \sigma \in \Sigma\}. \tag{2.9}$$

The integral $\int_0^\infty f(t)\, dt$ is called the *archimedean integral* of f. It enjoys the usual properties of the Riemann integral (see Exercise 2.5) but, among these, we will need only the monotonicity with respect to f in the sequel. For our purposes the most relevant property of the Archimedean integral is instead the continuity under monotonically nondecreasing sequences.

Proposition 2.8. *Let $f_n \uparrow f$, with $f_n : [0, +\infty) \to [0, +\infty]$ nonincreasing. Then*

$$\int_0^\infty f_n(t)\, dt \uparrow \int_0^\infty f(t)\, dt.$$

Proof. It is obvious that

$$\int_0^\infty f_n(t)\, dt \le \int_0^\infty f(t)\, dt.$$

To prove the converse inequality, fix $L < \int_0^\infty f(t)\, dt$. Then there exists $\sigma = \{t_1, \ldots, t_N\} \in \Sigma$ such that

$$\sum_{k=0}^{N-1} f(t_k)(t_{k+1} - t_k) > L.$$

Since for n large enough

$$\int_0^\infty f_n(t)\,dt \ge \sum_{k=0}^{N-1} f_n(t_{k+1})(t_{k+1}-t_k) > L,$$

letting $n \to \infty$ we find that

$$\sup_{n\in\mathbb{N}} \int_0^\infty f_n(t)\,dt \ge L.$$

This implies

$$\sup_{n\in\mathbb{N}} \int_0^\infty f_n(t)\,dt \ge \int_0^\infty f(t)\,dt$$

and the conclusion follows. $\qquad\square$

2.4.4. Integral of a nonnegative measurable function

We are given a measure space $(X, \mathscr{E}, \mu)$ and an extended nonnegative $\mathscr{E}$–measurable function φ. Having the identity (2.7) in mind, we define

$$\int_X \varphi\,d\mu := \int_0^\infty \mu(\{\varphi > t\})\,dt. \qquad (2.10)$$

Notice that the function $t \mapsto \mu(\{\varphi > t\}) \in [0, +\infty]$ is nonnegative and nonincreasing in $[0, +\infty)$, so that its archimedean integral is well defined and (2.10) extends, by the remarks made at the end of Section 2.4.2, the integral elementarily defined on simple functions. If the integral is finite we say that φ is μ–*integrable*.

It follows directly from the analogous properties of the archimedean integral that the integral so defined is monotone, *i.e.*

$$\varphi \ge \psi \implies \int_X \varphi\,d\mu \ge \int_X \psi\,d\mu.$$

Indeed, $\varphi \ge \psi$ implies $\{\varphi > t\} \supset \{\psi > t\}$ and $\mu(\{\varphi > t\}) \ge \mu(\{\psi > t\})$ for all $t > 0$. Furthermore, the integral is *invariant* under modifications of φ in μ–negligible sets, that is

$$\varphi = \psi\ \mu\text{–a.e. in } X \implies \int_X \varphi\,d\mu = \int_X \psi\,d\mu.$$

To show this fact it suffices to notice that $\varphi = \psi\ \mu$–a.e. in X implies that the sets $\{\varphi > t\}$ and $\{\psi > t\}$ differ in a μ–negligible set for all $t > 0$, therefore $\mu(\{\varphi > t\}) = \mu(\{\psi > t\})$ for all $t > 0$.

Let us prove the following basic *Markov inequality*.

Proposition 2.9. *For any $a \in (0, +\infty)$ we have*

$$\mu(\{\varphi \geq a\}) \leq \frac{1}{a} \int_X \varphi(x)\, d\mu(x). \tag{2.11}$$

Proof. For any $a \in (0, +\infty)$ we have, recalling the inclusion $\{\varphi \geq a\} \subset \{\varphi > t\}$ for any $t \in (0, a)$, that $\mu(\{\varphi > t\}) \geq \mu(\{\varphi \geq a\})$ for all $t \in (0, a)$. The monotonicity of the archimedean integral gives

$$\int_X \varphi(x)\, d\mu(x) = \int_0^\infty \mu(\{\varphi > t\})\, dt \geq \int_0^\infty \mathbb{1}_{(0,a)}(t)\mu(\{\varphi > t\})\, dt$$

$$\geq a\mu(\{\varphi \geq a\}). \qquad \square$$

The Markov inequality has some important consequences.

Proposition 2.10. *Let $\varphi : X \to [0, +\infty]$ be an extended $\mathcal{E}$–measurable function.*

(i) *If φ is μ–integrable then the set $\{\varphi = +\infty\}$ has μ–measure 0, that is, φ is finite μ–a.e. in X.*
(ii) *The integral of φ vanishes iff φ is equal to 0 μ–a.e. in X.*

Proof. (i) Since $\int_X \varphi\, d\mu < \infty$ we deduce from (2.11) that

$$\lim_{a \to +\infty} \mu(\{\varphi > a\}) = 0.$$

Since

$$\{\varphi = \infty\} = \bigcap_{n=1}^\infty \{\varphi > n\},$$

by applying the continuity along decreasing sequences in the space $(\{\varphi > 1\}$ (with finite μ measure) we obtain

$$\mu(\{\varphi = \infty\}) = \lim_{n \to +\infty} \mu(\{\varphi > n\}) = 0.$$

(ii) If $\int_X \varphi\, d\mu = 0$ we deduce from (2.11) that $\mu(\{\varphi > a\}) = 0$ for all $a > 0$. Since

$$\mu(\{\varphi > 0\}) = \lim_{n \to +\infty} \mu(\{\varphi > \frac{1}{n}\}) = 0,$$

the conclusion follows. The other implication follows by the invariance of the integral. $\qquad \square$

Proposition 2.11 (Monotone convergence). *Let (φ_n) be a nondecreasing sequence of extended nonnegative $\mathcal{E}$–measurable functions and set $\varphi(x) := \lim_{n \to \infty} \varphi_n(x)$ for any $x \in X$. Then*

$$\int_0^\infty \varphi_n(x)\, d\mu(x) \;\uparrow\; \int_0^\infty \varphi(x)\, d\mu(x).$$

Proof. It suffices to notice that $\mu(\{\varphi_n > t\}) \uparrow \mu(\{\varphi > t\})$ for all $t > 0$, and then to apply Proposition 2.8. $\qquad\square$

Now, by Proposition 2.4 we obtain the following important approximation property.

Proposition 2.12. *Let $\varphi : X \to [0, +\infty]$ be an extended $\mathcal{E}$–measurable function. Then there exist simple $\mathcal{E}$–measurable functions $\varphi_n : X \to [0, +\infty)$ such that $\varphi_n \uparrow \varphi$, so that*

$$\int_0^\infty \varphi_n(x)\, d\mu(x) \;\uparrow\; \int_0^\infty \varphi(x)\, d\mu(x).$$

Remark 2.13 (Construction of Lebesgue and Riemann integrals). Proposition 2.12 could be used as an alternative, and equivalent, definition of the Lebesgue integral: we can just define it as the supremum of the integral of minorant simple functions. This alternative definition is closer to the definitions of Archimedean integrals and of inner Riemann integral: the only (fundamental) difference is due to the choice of the family of "simple" functions. In all cases simple functions take finitely many values, but within the Lebesgue theory their level sets belong to a σ–algebra, and so the family of simple function is much richer, in comparison with the other theories.

We can now prove the additivity property of the integral.

Proposition 2.14. *Let $\varphi, \psi : X \to [0, \infty]$ be $\mathcal{E}$–measurable functions. Then*

$$\int_X (\varphi + \psi)\, d\mu = \int_X \varphi\, d\mu + \int_X \psi\, d\mu.$$

Proof. Let φ_n, ψ_n be simple functions with $\varphi_n \uparrow \varphi$ and $\psi_n \uparrow \psi$. Then, the additivity of the integral on simple functions gives

$$\int_X (\varphi_n + \psi_n)\, d\mu = \int_X \varphi_n\, d\mu + \int_X \psi_n\, d\mu.$$

We conclude passing to the limit as $n \to \infty$ and using the monotone convergence theorem. $\qquad\square$

The following Fatou's lemma, providing a semicontinuity property of the integral, is of basic importance.

Lemma 2.15 (Fatou). *Let $\varphi_n : X \to [0, +\infty]$ be extended $\mathcal{E}$–measurable functions. Then we have*

$$\int_X \liminf_{n\to\infty} \varphi_n(x)\, d\mu(x) \leq \liminf_{n\to\infty} \int_X \varphi_n(x)\, d\mu(x). \qquad (2.12)$$

Proof. Setting $\varphi(x) := \liminf_n \varphi_n(x)$, and $\psi_n(x) = \inf_{m \geq n} \varphi_m(x)$, we have that $\psi_n(x) \uparrow \varphi(x)$. Consequently, by the monotone convergence theorem,

$$\int_X \varphi(x)\, d\mu(x) = \lim_{n\to\infty} \int_X \psi_n(x)\, d\mu(x).$$

On the other hand

$$\int_X \psi_n(x)\, d\mu(x) \leq \int_X \varphi_n(x)\, d\mu(x),$$

so that

$$\int_X \varphi(x)\, d\mu(x) \leq \liminf_{n\to\infty} \int_X \varphi_n(x)\, d\mu(x). \qquad \square$$

In particular, if φ_n are pointwise converging to φ, we have

$$\int_X \varphi(x)\, d\mu(x) \leq \liminf_{n\to\infty} \int_X \varphi_n(x)\, d\mu(x).$$

2.5. Integral of functions with a variable sign

Let $\varphi \colon X \to \overline{\mathbb{R}}$ be an extended $\mathcal{E}$–measurable function. We say that φ is *μ–integrable* if both the positive part $\varphi^+(x) := \max\{\varphi(x), 0\}$ and the negative part $\varphi^-(x) := \max\{-\varphi(x), 0\}$ of φ are μ–integrable in X. As $\varphi = \varphi^+ - \varphi^-$, in this case it is natural to define

$$\int_X \varphi(x)\, d\mu(x) := \int_X \varphi^+(x)\, d\mu(x) - \int_X \varphi^-(x)\, d\mu(x).$$

As $|\varphi| = \varphi^+ + \varphi^-$, the additivity properties of the integral give that

$$\varphi \text{ is } \mu\text{–integrable} \qquad \text{if and only if} \qquad \int_X |\varphi|\, d\mu < \infty.$$

Let $\varphi \colon X \to \mathbb{R}$ and let $A \in \mathcal{E}$ be such that $\mathbb{1}_A \varphi$ is μ-integrable. We define also

$$\int_A \varphi(x)\, d\mu(x) := \int_X \mathbb{1}_A(x)\varphi(x)\, d\mu(x).$$

In the following proposition we summarize the main properties of the integral.

Proposition 2.16. *Let* φ, $\psi : X \to \overline{\mathbb{R}}$ *be* μ*–integrable functions.*

(i) *For any* α, $\beta \in \mathbb{R}$ *we have that* $\alpha\varphi + \beta\psi$ *is* μ*–integrable and*

$$\int_X (\alpha\varphi + \beta\psi)\, d\mu = \alpha \int_X \varphi\, d\mu + \beta \int_X \psi\, d\mu.$$

(ii) *If* $\varphi \le \psi$ *in* X *we have* $\displaystyle\int_X \varphi\, d\mu \le \int_X \psi\, d\mu.$

(iii) $\displaystyle\left| \int_X \varphi\, d\mu \right| \le \int_X |\varphi|\, d\mu.$

Proof. (i). Since $(-\varphi)^+ = \varphi^-$ and $(-\varphi)^- = \varphi^+$, we have $\int_X -\varphi\, d\mu = -\int_X \varphi\, d\mu$. So, possibly replacing φ by $-\varphi$ and ψ by $-\psi$ we can assume that $\alpha \ge 0$ and $\beta \ge 0$. We have

$$(\alpha\varphi + \beta\psi)^+ + \alpha\varphi^- + \beta\psi^- = (\alpha\varphi + \beta\psi)^- + \alpha\varphi^+ + \beta\psi^+,$$

so that we can integrate both sides and use the additivity on nonnegative functions to obtain

$$\int_X (\alpha\varphi + \beta\psi)^+\, d\mu + \alpha \int_X \varphi^-\, d\mu + \beta \int_X \psi^-\, d\mu$$
$$= \int_X (\alpha\varphi + \beta\psi)^-\, d\mu + \alpha \int_X \varphi^+\, d\mu + \beta \int_X \psi^+\, d\mu.$$

Rearranging terms we obtain (i).

(ii). It follows by the monotonicity of the integral on nonnegative functions and from the inequalities $\varphi^+ \le \psi^+$ and $\varphi^- \ge \psi^-$.

(iii). Since $-|\varphi| \le \varphi \le |\varphi|$ the conclusion follows from (ii). $\qquad\square$

Another consequence of the additivity property of the integral is the additivity of the real-valued map

$$A \in \mathscr{E} \mapsto \int_A \varphi\, d\mu$$

whenever φ is μ–integrable. We will see in the next section that, as a consequence of the dominated convergence theorem, this map is even σ–additive.

2.6. Convergence of integrals

In this section we study the problem of commuting limit and integral; we have already seen that this can be done in some particular cases, as when the functions are nonnegative and monotonically converge to their supremum, and now we investigate some more general cases, relevant for the applications.

Proposition 2.17 (Lebesgue dominated convergence theorem). *Let (φ_n) be a sequence of $\mathscr{E}$–measurable functions pointwise converging to φ. Assume that there exists a nonnegative μ–integrable function ψ such that*

$$|\varphi_n(x)| \le \psi(x) \qquad \forall x \in X, \ n \in \mathbb{N}.$$

Then the functions φ_n and the function φ are μ–integrable and

$$\lim_{n \to \infty} \int_X \varphi_n \, d\mu = \int_X \varphi \, d\mu.$$

Proof. Passing to the limit as $n \to \infty$ we obtain that φ is $\mathscr{E}$–measurable and $|\varphi| \le \psi$ in X. In particular φ is μ–integrable. Since $\varphi + \psi$ is nonnegative, by the Fatou lemma we have

$$\int_X (\varphi + \psi) \, d\mu \le \liminf_{n \to \infty} \int_X (\varphi_n + \psi) \, d\mu.$$

Consequently,

$$\int_X \varphi \, d\mu \le \liminf_{n \to \infty} \int_X \varphi_n \, d\mu. \tag{2.13}$$

In a similar way we have

$$\int_X (\psi - \varphi) \, d\mu \le \liminf_{n \to \infty} \int_X (\psi - \varphi_n) \, d\mu.$$

Consequently,

$$\int_X \varphi \, d\mu \ge \limsup_{n \to \infty} \int_X \varphi_n \, d\mu. \tag{2.14}$$

Now the conclusion follows by (2.13) and (2.14). $\qquad\square$

An important consequence of the dominated convergence theorem is the *absolute continuity* property of the integral of μ–integrable functions φ:

for any $\varepsilon > 0$ there exists $\delta > 0$ such that $\mu(A) < \delta \implies \int_A |\varphi| \, d\mu < \varepsilon.$

$$\tag{2.15}$$

The proof of this property is sketched in Exercise 2.9.

2.6.1. Uniform integrability and Vitali convergence theorem

In this subsection we assume for simplicity that the measure μ is finite. A family $\{\varphi_i\}_{i\in I}$ of $\overline{\mathbb{R}}$–valued μ–integrable functions is said to be μ–*uniformly integrable* if

$$\lim_{\mu(A)\to 0} \int_A |\varphi_i(x)|\, d\mu(x) = 0, \quad \text{uniformly in } i \in I.$$

This means that for any $\varepsilon > 0$ there exists $\delta > 0$ such that

$$\mu(A) < \delta \implies \int_A |\varphi_i(x)|\, d\mu(x) \le \varepsilon \quad \forall\, i \in I.$$

This property obviously extends from single functions to families of functions the absolute continuity property of the integral.

Notice that any family $\{\varphi_i\}_{i\in I}$ dominated by a single μ–integrable function φ (*i.e.* such that $|\varphi_i| \le |\varphi|$ for any $i \in I$) is obviously μ–uniformly integrable. Taking this remark into account, we are going to to prove the following extension of the dominated convergence theorem, known as *Vitali Theorem*.

Theorem 2.18 (Vitali). *Assume that μ is a finite measure and let (φ_n) be a μ–uniformly integrable sequence of functions pointwise converging to a real valued function φ. Then φ is μ–integrable and*

$$\lim_{n\to\infty} \int_X \varphi_n\, d\mu = \int_X \varphi\, d\mu.$$

To prove the Vitali theorem we need the following Egorov Lemma.

Lemma 2.19 (Egorov). *Assume that μ is a finite measure and let (φ_n) be a sequence of $\mathscr{E}$–measurable functions pointwise converging to a real valued function φ. Then for any $\delta > 0$ there exists a set $A_\delta \in \mathscr{E}$ such that $\mu(A_\delta) < \delta$ and $\varphi_n \to \varphi$ uniformly in $X \setminus A_\delta$.*

Proof. For any integer $m \ge 1$ we write X as the increasing union of the sets $B_{n,m}$, where

$$B_{n,m} := \left\{ x \in X \; : \; |\varphi_i(x) - \varphi(x)| < \frac{1}{m} \quad \forall i \ge n \right\}.$$

Since μ is finite there exists $n(m)$ such that $\mu(B_{n(m),m}) > \mu(X) - 2^{-m}\delta$. We denote by A_δ the union of $X \setminus B_{n(m),m}$, so that

$$\mu(A_\delta) \le \sum_{m=1}^{\infty} \mu(X \setminus B_{n(m),m}) < \sum_{m=1}^{\infty} \frac{\delta}{2^m} = \delta.$$

Now, given any $\varepsilon > 0$, we can choose $m > 1/\varepsilon$ to obtain that

$$|\varphi_n(x) - \varphi(x)| \leq \frac{1}{m} < \varepsilon \quad \text{for all } x \in B_{n(m),m}, n \geq n(m).$$

As $X \setminus A_\delta \subset B_{n(m),m}$, this proves the uniform convergence of φ_n to φ on $X \setminus A_\delta$. $\qquad\qquad\square$

Proof of the Vitali Theorem. Fix $\varepsilon > 0$ and find $\delta > 0$ such that $\int_A |\varphi_n| \, d\mu < \varepsilon$ whenever $\mu(A) < \delta$. Again, Fatou's Lemma yields that $\int_A |\varphi| \, d\mu \leq \varepsilon$ whenever $\mu(A) < \delta$.

Assume now that A is given by Egorov Lemma, so that $\varphi_n \to \varphi$ uniformly on $X \setminus A$. Then, writing

$$\int_X (\varphi - \varphi_n) \, d\mu = \int_{X \setminus A} (\varphi - \varphi_n) \, d\mu + \int_A (\varphi - \varphi_n) \, d\mu$$

and using the fact that $\lim_n \sup_{X \setminus A} |\varphi_n - \varphi| = 0$ we obtain

$$\left| \int_X (\varphi - \varphi_n) \, d\mu \right| \leq 3\varepsilon$$

for n large enough. The statement follows letting $\varepsilon \downarrow 0$. $\qquad\qquad\square$

2.7. A characterization of Riemann integrable functions

The integrals $\int_J f \, d\lambda$, with $J = [a, b]$ closed interval of the real line and λ equal to the Lebesgue measure in $\mathbb{R}$, are traditionally denoted with the classical notation $\int_a^b f \, dx$ or with $\int_J f \, dx$. This is due to the fact that Riemann's and Lebesgue's integral coincide on the class of Riemann's integrable functions.

We denote by $I_*(f)$ and $I^*(f)$ the upper and lower Riemann integral of f respectively, the former defined by taking the supremum of the sums $\sum_1^{n-1} a_i(t_{i+1} - t_i)$ in correspondence of all *step functions*

$$h = \sum_{i=1}^{n-1} a_i \mathbb{1}_{[t_i, t_{i+1})} \leq f \qquad\qquad a = t_1 < \cdots < t_n = b, \qquad (2.16)$$

and the latter considering the infimum in correspondence of all step functions $h \geq f$. We denote by $I(f)$ the Riemann integral, equal to the upper and lower integral whenever the two integrals coincide.

As the Lebesgue integral of the function h in (2.19) coincides with $\sum_i^{n-1} a_i(t_{i+1} - t_i)$, we have

$$\int_J g \, d\lambda = I(g) \qquad \text{for any step function } g : J \to \mathbb{R}.$$

Now, if $f : J \to \mathbb{R}$ is continuous, we can choose a uniformly bounded sequence of step functions g_h converging pointwise to f (for instance splitting J into i equal intervals $[x_i, x_{i+1}[$ and setting $a_i = \min_{[x_i, x_{i+1}]} f$) whose Riemann integrals converge to $I(f)$. Therefore, passing to the limit in the identity above with $g = g_h$, and using the dominated convergence theorem we get

$$\int_J f \, d\lambda = I(f) \qquad \text{for any continuous function } f : J \to \mathbb{R}.$$

We are going to generalize this fact, providing a full characterization, within the Lebesgue theory, of Riemman's integrable functions.

Theorem 2.20. *Let* $f : J = [a, b] \to \mathbb{R}$ *be a bounded function. Then* f *is Riemann integrable if and only if the set of its discontinuity points is Lebesgue negligible. If this is the case, we have that* f *is* $\mathscr{B}(J)_{\lambda}$–*measurable and*

$$\int_J f \, d\lambda = I(f). \tag{2.17}$$

Proof. Let

$$\begin{cases} f_*(x) := \inf \left\{ \liminf_{h \to \infty} f(x_h) : \ x_h \to x \right\} \\[3mm] f^*(x) := \sup \left\{ \limsup_{h \to \infty} f(x_h) : \ x_h \to x \right\}. \end{cases} \tag{2.18}$$

It is not hard to show (see Exercise 2.6 and Exercise 2.7) that f_* is lower semicontinuous and f^* is upper semicontinuous, therefore both f_* and f^* are Borel functions.

We are going to show that $I_*(f) = \int_J f_* \, d\lambda$ and $I^*(f) = \int_J f^* \, d\lambda$. These two equalities yield the conclusion, as f is continuous at λ–a.e. point in J iff $f^* - f_* = 0$ λ–a.e. in J, and this holds iff (because $f^* - f_* \geq 0$)

$$\int_J (f^* - f_*) \, d\lambda = 0.$$

Furthermore, if the set of discontinuity points of f is λ–negligible, the Borel function f^* differs from f only in a λ–negligible set, thus f is $\mathscr{B}(J)_{\lambda}$–measurable (because $\{f > t\}$ differs from the Borel set $\{f^* > t\}$ only in a λ–negligible set, see also Exercise 2.4) and its integral coincides with $\int_J f^* \, d\lambda = \int_J f_* \, d\lambda$; this leads to (2.17).

Since $I^*(f) = -I_*(-f)$ and $f^* = -(-f)_*$, we need only to prove the first of the two equalities, *i.e.*

$$\int_J f_* \, d\lambda = I_*(f). \tag{2.19}$$

In order to check the inequality $\leq$ in (2.19) we apply Exercise 2.11, finding a sequence of continuous functions $g_h \uparrow f_* \leq f$ and obtaining, thanks to the monotone convergence theorem,

$$\int_J f_* \, d\lambda = \sup_{h \in \mathbb{N}} \int_J g_h \, d\lambda = \sup_{h \in \mathbb{N}} I(g_h) = \sup_{h \in \mathbb{N}} I_*(g_h) \leq I_*(f).$$

In order to prove $\geq$ in (2.19) we fix a step function $h \leq f$ in $[a, b)$ as in (2.16) and we notice that $f \geq a_i = h$ in (t_i, t_{i+1}) implies $f_* \geq a_i$ in the same interval. Hence $f_* \geq h$ in $J \setminus \{t_1, \ldots, t_n\}$ and, being the set of the t_i's Lebesgue negligible, we have

$$\int_J f_* \, d\lambda \geq \int_J h \, d\lambda = I(h).$$

Since h is arbitrary the inequality is achieved. $\qquad\square$

Exercises

2.1 Show that any of the conditions listed below is equivalent to the $\mathscr{E}$–measurability of $\varphi : X \to \mathbb{R}$.

- (i) $\varphi^{-1}((-\infty, t]) \subset \mathscr{E}$ for all $t \in \mathbb{R}$;
- (ii) $\varphi^{-1}((-\infty, t)) \subset \mathscr{E}$ for all $t \in \mathbb{R}$;
- (iii) $\varphi^{-1}([a, b]) \subset \mathscr{E}$ for all $a, b \in \mathbb{R}$;
- (iv) $\varphi^{-1}([a, b)) \subset \mathscr{E}$ for all $a, b \in \mathbb{R}$;
- (v) $\varphi^{-1}((a, b)) \subset \mathscr{E}$ for all $a, b \in \mathbb{R}$.

2.2 Let $\varphi, \psi : X \to \mathbb{R}$ be $\mathscr{E}$–measurable. Show that $\varphi + \psi$ and $\varphi\psi$ are $\mathscr{E}$–measurable. *Hint:* prove that

$$\{\varphi + \psi < t\} = \bigcup_{r \in \mathbb{Q}} [\{\varphi < r\} \cap \{\psi < t - r\}]$$

and

$$\{\varphi^2 > a\} = \{\varphi > \sqrt{a}\} \cup \{\varphi < -\sqrt{a}\}, \quad a \geq 0.$$

2.3 Let us define a distance d in $\overline{\mathbb{R}}$ by

$$d(x, y) := |\arctan x - \arctan y|$$

where, by convention, $\arctan(\pm\infty) = \pm\pi/2$.

(i) Show that $(\overline{\mathbb{R}}, d)$ is a compact metric space (the so-called *compactification of* $\mathbb{R}$) and that $A \subset \mathbb{R}$ is open relative to the Euclidean distance if, and only if, it is open relative to d;

(ii) use (i) to show that, given a measurable space $(X, \mathscr{E})$, $f : X \to \overline{\mathbb{R}}$ is $\mathscr{E}$–measurable according to (2.2) if and only if it is measurable between $\mathscr{E}$ and the Borel σ–algebra of $(\overline{\mathbb{R}}, d)$.

2.4 Let $(X, \mathscr{E}, \mu)$ be a measure space and let $\mathscr{E}_\mu$ be the completion of $\mathscr{E}$ induced by μ. Show that $f : X \to \mathbb{R}$ is $\mathscr{E}_\mu$–measurable iff there exists a $\mathscr{E}$–measurable function g such that $\{f \neq g\}$ is contained in a μ–negligible set of $\mathscr{E}$.

2.5 Let us define I_f as in (2.8) and let us endow Σ with the usual partial ordering $\sigma = \{t_1, \ldots, t_N\} \leq \zeta = \{s_1, \ldots, s_M\}$ if and only if $\sigma \subset \zeta$. Show that $\sigma \mapsto I_f(\sigma)$ is nondecreasing. Use this fact to show that $f \mapsto \int_0^\infty f(t)\, dt$ is additive.

2.6 Let $f : \mathbb{R} \to \mathbb{R}$ be a function. Show that the functions f_*, f^* defined in (2.18) are respectively lower semicontinuous and upper semicontinuous.

2.7 Let $f : \mathbb{R} \to \mathbb{R}$ be a bounded function. Using Exercise 2.6 show that $\{f_* \leq t\}$ and $\{f^* \geq t\}$ are closed for all $t \in \mathbb{R}$. In particular deduce that

$$\Sigma = \{x \in \mathbb{R} \ : \ f \text{ is continuous at } x\}$$

belongs to $\mathscr{B}(\mathbb{R})$.

2.8 Let $(a_n) \subset (0, \infty)$ with

$$\sum_{i=0}^\infty a_i = \infty, \qquad \lim_{i \to \infty} a_i = 0.$$

Show that for any $\varphi : X \to [0, +\infty]$ $\mathscr{E}$–measurable there exist $A_i \in \mathscr{E}$ such that $\varphi = \sum_i a_i \mathbb{1}_{A_i}$. *Hint:* set $\varphi_0 := \varphi$, $A_0 := \{\varphi \geq a_0\}$ and $\varphi_1 := \varphi_0 - a_0 \mathbb{1}_{A_0} \geq 0$. Then, set $A_1 := \{\varphi_1 \geq a_1\}$ and $\varphi_2 := \varphi_0 - a_1 \mathbb{1}_{A_1}$ and so on.

2.9 Let $\varphi : X \to \overline{\mathbb{R}}$ be μ–integrable. Show that the property (2.15) holds. *Hint:* assume by contradiction its failure for some $\varepsilon > 0$ and find A_i with $\mu(A_i) < 2^{-i}$ and $\int_{A_i} |\varphi|\, d\mu \geq \varepsilon$. Then, notice that $B := \limsup_i A_i$ is μ–negligible, consider

$$B_n := \bigcup_{i \geq n} A_i \setminus B \downarrow \varnothing$$

and apply the dominated convergence theorem.

2.10 Prove that if $\varphi_n \to \varphi$ in $L^1(\Omega, \mathscr{E}, \mu)$, then (φ_n) is μ–uniformly integrable. In addition, find a space $(X, \mathscr{E}, \mu)$ and a sequence (φ_n) that is μ–uniformly integrable, for which there is no $g \in L^1(X, \mathscr{E}, \mu)$ satisfying $|\varphi_n| \leq g$ for all $n \in \mathbb{N}$.

2.11 Let (X, d) be a metric space and let $g : X \to [0, \infty]$ be lower semicontinuous and not identically equal to ∞. For any $\lambda > 0$ define

$$g_\lambda(x) := \inf_{y \in X} \{g(y) + \lambda d(x, y)\}.$$

Check that:

(a) $|g_\lambda(x) - g_\lambda(x')| \le \lambda d(x, x')$ for all $x,\, x' \in X$;
(b) $g_\lambda \uparrow g$ as $\lambda \uparrow \infty$.

2.12 Let $f : \mathbb{R}^2 \to \mathbb{R}$ be satisfying the following two properties:

(i) $x \mapsto f(x, y)$ is continuous in $\mathbb{R}$ for all $y \in \mathbb{R}$;
(ii) $y \mapsto f(x, y)$ is continuous in $\mathbb{R}$ for all $x \in \mathbb{R}$.

Show that f is a Borel function. *Hint:* first reduce to the case when f is bounded. Then, for $\varepsilon > 0$ consider the functions

$$f_\varepsilon(x, y) := \frac{1}{2\varepsilon} \int_{x-\varepsilon}^{x+\varepsilon} f(x', y)\, dx',$$

proving that f_ε are continuous and $f_\varepsilon \to f$ as $\varepsilon \downarrow 0$.

Chapter 3
Spaces of integrable functions

This chapter is devoted to the properties of the so-called L^p spaces, the spaces of measurable functions whose p-th power is integrable. Throughout the chapter a measure space $(X, \mathscr{E}, \mu)$ will be fixed.

3.1. Spaces $\mathscr{L}^p(X, \mathscr{E}, \mu)$ and $L^p(X, \mathscr{E}, \mu)$

Let Y be a real vector space. We recall that a *norm* $\| \cdot \|$ on Y is a non-negative map defined on Y satisfying:

  (i) $\|y\| = 0$ if and only if $y = 0$;
 (ii) $\|\alpha y\| = |\alpha| \, \|y\|$ for all $\alpha \in \mathbb{R}$ and $y \in Y$;
(iii) $\|y_1 + y_2\| \le \|y_1\| + \|y_2\|$ for all $y_1, \, y_2 \in Y$.

The space Y, endowed with the norm $\| \cdot \|$, is called a *normed space*. Y is also a metric space when endowed with the distance $d(y_1, y_2) = \|y_1 - y_2\|$ (the triangle inequality is a direct consequence of (iii)). If (Y, d) is a *complete* metric space, we say that $(Y, \| \cdot \|)$ is a *Banach space*.

We denote by $\mathscr{L}^1(X, \mathscr{E}, \mu)$ the real vector space of all μ–integrable functions on $(X, \mathscr{E})$. We define

$$\|\varphi\|_1 := \int_X |\varphi(x)| \, d\mu(x), \qquad \varphi \in \mathscr{L}^1(X, \mathscr{E}, \mu).$$

We have clearly

$$\|\alpha\varphi\|_1 = |\alpha| \, \|\varphi\|_1 \qquad \forall \alpha \in \mathbb{R}, \ \forall \varphi \in \mathscr{L}^1(X, \mathscr{E}, \mu),$$

and

$$\|\varphi + \psi\|_1 \le \|\varphi\|_1 + \|\psi\|_1 \qquad \forall \varphi, \, \psi \in \mathscr{L}^1(X, \mathscr{E}, \mu),$$

so that conditions (ii) and (iii) in the definition of the norm are fulfilled. However, $\| \cdot \|_1$ is not a norm in general, since $\|\varphi\|_1 = 0$ if and only if $\varphi = 0$ μ–a.e. in X, so (i) fails.

Then, we can consider the following equivalence relation $\mathscr{R}$ on $\mathscr{L}^1(X, \mathscr{E}, \mu)$,

$$\varphi \sim \psi \quad \Longleftrightarrow \quad \varphi = \psi \ \mu\text{–a.e. in } X \tag{3.1}$$

and denote by $L^1(X, \mathscr{E}, \mu)$ the quotient space of $\mathscr{L}^1(X, \mathscr{E}, \mu)$ with respect to $\mathscr{R}$. In other words, $L^1(X, \mathscr{E}, \mu)$ is the quotient vector space of $\mathscr{L}^1(X, \mathscr{E}, \mu)$ with respect to the vector subspace made by functions vanishing μ–a.e. in X.

For any $\varphi \in \mathscr{L}^1(X, \mathscr{E}, \mu)$ we denote by $\tilde{\varphi}$ the equivalence class determined by φ and we set

$$\tilde{\varphi} + \tilde{\psi} := \widetilde{\varphi + \psi}, \qquad \alpha\tilde{\varphi} := \widetilde{\alpha\varphi}. \tag{3.2}$$

It is easily seen that these definitions do no depend on the choice of representatives in the equivalence class, and endow $L^1(X, \mathscr{E}, \mu)$ with the structure of a real vector space, whose origin is the equivalence class of functions vanishing μ–a.e. in X. Furthermore, setting

$$\|\tilde{\varphi}\|_1 = \|\varphi\|_1, \qquad \tilde{\varphi} \in L^1(X, \mathscr{E}, \mu),$$

it is also easy to see that this definition does not depend on the particular element φ chosen in $\tilde{\varphi}$, and that (ii), (iii) still hold. Now, if $\|\tilde{\varphi}\|_1 = 0$ we have that the integral of $|\varphi|$ is zero, and therefore $\tilde{\varphi} = 0$. Therefore $L^1(X, \mathscr{E}, \mu)$, endowed with the norm $\| \cdot \|_1$, is a normed space.

To simplify the notation typically $\tilde{\varphi}$ is identified with φ whenever the formula does not depend on the choice of the function in the equivalence class: for instance, quantities as $\mu(\{\varphi > t\})$ or $\int_X \varphi \, d\mu$ have this independence, as well as most statements and results in Measure Theory and Probability, so this slight abuse of notation is justified. It should be noted, however, that formulas like $\varphi(\bar{x}) = 0$, for some fixed $\bar{x} \in X$, do not make sense in $L^1(X, \mathscr{E}, \mu)$, since they depend on the representative chosen (unless $\mu(\{\bar{x}\}) > 0$).

More generally, if an exponent $p \in (0, \infty)$ is given, we can apply a similar construction to the space

$$\mathscr{L}^p(X, \mathscr{E}, \mu) := \left\{ \varphi \ : \ \varphi \text{ is } \mathscr{E}\text{–measurable and } \int_X |\varphi|^p \, d\mu < \infty \right\}.$$

Since $|x + y|^p \leq |x|^p + |y|^p$ if $p \leq 1$, and $|x + y|^p \leq 2^{p-1}(|x|^p + |y|^p)$ if $p \geq 1$, it turns out that $\mathscr{L}^p(X, \mathscr{E}, \mu)$ is a vector space, and we shall denote by $L^p(X, \mathscr{E}, \mu)$ the quotient vector space, with respect to the equivalence relation (3.1). Still we can define the sum and product by a real number as in (3.2), to obtain that $L^p(X, \mathscr{E}, \mu)$ has the structure of a real vector space. The case $p = 2$ is particularly relevant for the theory, as we will see.

Sometimes we will omit either $\mathscr{E}$ or μ, writing $L^p(X, \mu)$ or even $L^p(X)$. This typically happens when (X, d) is a metric space, and $\mathscr{E}$ is the Borel σ-algebra, or when $X \subset \mathbb{R}$ and μ is the Lebesgue measure.

3.2. The L^p norm

For any $\varphi \in L^p(X, \mathscr{E}, \mu)$ we define

$$\|\varphi\|_p := \left(\int_X |\varphi|^p \, d\mu \right)^{1/p}.$$

We are going to show that $\| \cdot \|_p$ is a norm for any $p \in [1, +\infty)$. Notice that we already checked this fact when $p = 1$, and that the homogeneity condition (ii) trivially holds, whatever the value of p is. Furthermore, condition (i) holds precisely because $L^p(X, \mathscr{E}, \mu)$ consists, strictly speaking, of equivalence classes induced by (3.1). So, the only condition that needs to be checked is the subadditivity condition (ii), and in the sequel we can assume $p > 1$.

The concept of *Legendre transform* will be useful. Let $f : \mathbb{R} \to \mathbb{R}$ be a function; we define its Legendre transform $f^* : \mathbb{R} \to \mathbb{R} \cup \{+\infty\}$ by

$$f^*(y) = \sup_{x \in \mathbb{R}} \{xy - f(x)\}, \quad y \in \mathbb{R}.$$

Then the following inequality clearly holds:

$$xy \leq f(x) + f^*(y) \qquad \forall x, y \in \mathbb{R}, \tag{3.3}$$

and actually f^* could be equivalently defined as the smallest function with this property.

Example 3.1. Let $p > 1$ and let

$$f(x) = \begin{cases} \dfrac{x^p}{p} & \text{if } x \geq 0, \\[2mm] 0 & \text{if } x < 0. \end{cases}$$

Then, by an elementary computation, we find that

$$f^*(y) = \begin{cases} \dfrac{y^q}{q} & \text{if } y \geq 0, \\[2mm] +\infty & \text{if } y < 0, \end{cases}$$

where $q = p/(p-1)$ (equivalently, $\frac{1}{p} + \frac{1}{q} = 1$). Consequently, the following inequality, known as *Young inequality*, holds:

$$xy \leq \frac{x^p}{p} + \frac{y^q}{q}, \qquad x, y \geq 0. \tag{3.4}$$

Motivated by the previous example, we say that p and q are *dual* (or conjugate) exponents if $\frac{1}{p} + \frac{1}{q} = 1$, *i.e.* $q = p/(p-1)$. The duality relation is symmetric in $(1, +\infty)$, and obviously 2 is self-dual.

Example 3.2. Let $f(x) = e^x$, $x \in \mathbb{R}$. Then

$$f^*(y) := \sup_{x \in \mathbb{R}}\{xy - e^x\} = \begin{cases} +\infty & \text{if } y < 0, \\ 0 & \text{if } y = 0, \\ y \log y - y & \text{if } y > 0. \end{cases}$$

Consequently, the following inequality holds:

$$xy \leq e^x + y \log y - y, \qquad x, y \geq 0. \tag{3.5}$$

3.2.1. Hölder and Minkowski inequalities

Proposition 3.3 (Hölder inequality). *Assume that $\varphi \in L^p(X, \mathcal{E}, \mu)$ and $\psi \in L^q(X, \mathcal{E}, \mu)$, with p and q dual exponents in $(1, +\infty)$. Then $\varphi\psi \in L^1(X, \mathcal{E}, \mu)$ and*

$$\|\varphi\psi\|_1 \leq \|\varphi\|_p \|\psi\|_q. \tag{3.6}$$

Proof. If either $\|\varphi\|_p = 0$ or $\|\psi\|_q = 0$ then one of the two functions vanishes μ–a.e. in X, hence $\varphi\psi$ vanishes μ–a.e. and the inequality is trivial. If both $\|\varphi\|_p$ and $\|\psi\|_q$ are strictly positive, by the 1–homogeneity of the both sides in (3.6) with respect to φ and ψ, we can assume with no loss of generality that the two norms are equal to 1.

Now we apply (3.4) to $|\varphi(x)|$ and $|\psi(x)|$ to obtain

$$|\varphi(x)\psi(x)| \leq \frac{|\varphi(x)|^p}{p} + \frac{|\psi(x)|^q}{q}.$$

Integrating over X with respect to μ yields

$$\int_X |\varphi(x)\psi(x)| \, d\mu(x) \leq \frac{1}{p} + \frac{1}{q} = 1. \qquad \square$$

A particular case of the Hölder inequality is

$$\left|\int_X \varphi(x)\psi(x) \, d\mu(x)\right| \leq \left(\int_X \varphi^2(x) \, d\mu(x)\right)^{1/2} \left(\int_X \psi^2(x) \, d\mu(x)\right)^{1/2}.$$

It also follows, as we shall see, from the Cauchy-Schwarz inequality of scalar products.

Proposition 3.4 (Minkowski inequality). *Assume that $p \in [1, +\infty)$ and $\varphi, \psi \in L^p(X, \mathscr{E}, \mu)$. Then $\varphi + \psi \in L^p(X, \mathscr{E}, \mu)$ and*

$$\|\varphi + \psi\|_p \leq \|\varphi\|_p + \|\psi\|_p. \tag{3.7}$$

Proof. The cases $p = 1$ is obvious. Assume that $p \in (1, +\infty)$. Then we have

$$\int_X |\varphi + \psi|^p \, d\mu \leq \int_X |\varphi + \psi|^{p-1} |\varphi| \, d\mu + \int_X |\varphi + \psi|^{p-1} |\psi| \, d\mu.$$

Since $|\varphi + \psi|^{p-1} \in L^q(X, \mathscr{E}, \mu)$ where $q = p/(p-1)$, using the Hölder inequality we find that

$$\int_X |\varphi + \psi|^p \, d\mu \leq \left(\int_X |\varphi + \psi|^p \, d\mu \right)^{1/q} (\|\varphi\|_p + \|\psi\|_p),$$

and the conclusion follows. $\square$

By the previous proposition it follows that $\|\cdot\|_p$ is a norm on $L^p(X, \mathscr{E}, \mu)$.

3.3. Convergence in $L^p(X, \mathscr{E}, \mu)$ and completeness

We have seen in the previous section that $L^p(X, \mathscr{E}, \mu)$ is a normed space for all $p \in [1, +\infty)$. In this section we prove some properties of the convergence in these spaces, obtaining as a byproduct the following result.

Theorem 3.5. *$L^p(X, \mathscr{E}, \mu)$ is a Banach space for any $p \in [1, +\infty)$.*

This theorem will be a direct consequence of the following proposition, that provides also a relation between convergence in L^p and convergence μ–a.e. in X.

Proposition 3.6. *Let $p \in [1, +\infty)$ and let (φ_n) be a Cauchy sequence in $L^p(X, \mathscr{E}, \mu)$. Then:*

(i) *there exists a subsequence $(\varphi_{n(k)})$ converging μ–a.e. to a function φ in $L^p(X, \mathscr{E}, \mu)$;*

(ii) *(φ_n) is converging to φ in $L^p(X, \mathscr{E}, \mu)$, so that $L^p(X, \mathscr{E}, \mu)$ is a Banach space.*

Proof. Let (φ_n) be a Cauchy sequence in $L^p(X, \mathscr{E}, \mu)$. Choose a subsequence $(\varphi_{n(k)})$ such that

$$\|\varphi_{n(k+1)} - \varphi_{n(k)}\|_p < 2^{-k} \qquad \forall k \in \mathbb{N}.$$

Next, set

$$g(x) := \sum_{k=0}^{\infty} |\varphi_{n(k+1)}(x) - \varphi_{n(k)}(x)|, \qquad x \in X.$$

By the monotone convergence theorem and the subadditivity of the L^p norm it follows that

$$\left(\int_X g^p(x)\, d\mu(x) \right)^{1/p} = \lim_{N \to \infty} \left(\int_X \left| \sum_{k=0}^{N-1} |\varphi_{n(k+1)} - \varphi_{n(k)}| \right|^p d\mu \right)^{1/p}$$

$$\leq \lim_{N \to \infty} \sum_{k=0}^{N-1} 2^{-k} = 2 < \infty.$$

Therefore, g is finite μ–a.e., that is, there exists $B \in \mathscr{E}$ such that $\mu(B) = 0$ and $g(x) < \infty$ for all $x \in B^c$. Set now

$$\varphi(x) := \varphi_{n(0)}(x) + \sum_{k=0}^{\infty} (\varphi_{n(k+1)}(x) - \varphi_{n(k)}(x)), \qquad x \in B^c.$$

The series above is absolutely convergent for any $x \in B^c$; moreover, replacing the series in the definition of φ by the finite sum $\sum_0^{N-1} (\varphi_{n(k+1)}(x) - \varphi_{n(k)}(x))$ we obtain $\varphi(x) = \lim_k \varphi_{n(k)}(x)$. Therefore, if we define (for instance) $\varphi = 0$ on the μ–negligible set B, we obtain that $\varphi_{n(k)} \to \varphi$ μ–a.e. on X.

The inequality $|\varphi| \leq |\varphi_{n(0)}| + g$ gives that $|\varphi|^p$ is μ–integrable, so that $\varphi \in L^p(X, \mathscr{E}, \mu)$. So, (i) is proved.

In order to prove (ii), we first claim that $\varphi_{n(k)} \to \varphi$ in $L^p(X, \mathscr{E}, \mu)$ as $k \to \infty$. In fact, since

$$|\varphi(x) - \varphi_{n(h)}(x)| \leq \sum_{k=h}^{\infty} |\varphi_{n(k+1)}(x) - \varphi_{n(k)}(x)|, \quad x \in X,$$

we have, again by monotone convergence and subadditivity of the norm,

$$\left(\int_X |\varphi(x) - \varphi_{n(h)}(x)|^p\, d\mu(x) \right)^{1/p}$$

$$\leq \sum_{k=h}^{\infty} \left(\int_X |\varphi_{n(k+1)}(x) - \varphi_{n(k)}(x)|^p\, d\mu(x) \right)^{1/p} \leq \sum_{k=h}^{\infty} 2^{-k},$$

and the claim follows.

Since (φ_n) is Cauchy, for any $\varepsilon > 0$ there exists $n_\varepsilon \in \mathbb{N}$ such that

$$n, m > n_\varepsilon \quad \Longrightarrow \quad \|\varphi_n - \varphi_m\|_p < \varepsilon.$$

Now choose $k \in \mathbb{N}$ such that $n(k) > n_\varepsilon$ and $\|\varphi - \varphi_{n(k)}\|_p < \varepsilon$. For any $n > n_\varepsilon$ we have

$$\|\varphi - \varphi_n\|_p \leq \|\varphi - \varphi_{n(k)}\|_p + \|\varphi_{n(k)} - \varphi_n\|_p \leq 2\varepsilon. \qquad \square$$

Remark 3.7 (L^p convergence versus μ–a.e. convergence).   The argument used in the previous proof applies also to converging sequences (as these sequences are obviously Cauchy), and proves that any sequence (φ_n) strongly converging to φ in $L^p(X,\mathscr{E}, \mu)$ admits a subsequence $(\varphi_{n(k)})$ converging μ–a.e. to φ: precisely, this happens whenever

$$\sum_0^\infty \|\varphi_{n(k+1)} - \varphi_{n(k)}\|_p < \infty.$$

In general, however, convergence in L^p does not imply convergence μ–a.e.: the functions

$$\begin{cases} \varphi_0 = \mathbb{1}_{[0,1]} \\ \varphi_1 = \mathbb{1}_{[0,1/2]}, \;\; \varphi_2 = \mathbb{1}_{[1/2,1]} \\ \varphi_3 = \mathbb{1}_{[0,1/3]}, \;\; \varphi_4 = \mathbb{1}_{[1/3,2/3]}, \;\; \varphi_5 = \mathbb{1}_{[2/3,1]} \\ \cdots \end{cases}$$

converge to 0 in $L^p(0, 1)$, but are nowhere pointwise converging.

The previous remark shows that we can expect to infer pointwise convergence from convergence in L^p only modulo the extraction of a subsequence. Now, we ask ourselves about the converse implication: given a sequence (φ_n) in $L^p(X, \mathscr{E}, \mu)$ pointwise converging to a function $\varphi \in L^p(X, \mathscr{E}, \mu)$, we want to find conditions ensuring the convergence of (φ_n) to φ in $L^p(X, \mathscr{E}, \mu)$. This is not true in general, as the following example shows.

Example 3.8. *Let $X = [0, 1]$, $\mathscr{E} = \mathscr{B}([0, 1])$ and let $\mu = \lambda$ be the Lebesgue measure. Set*

$$\varphi_n(x) = \begin{cases} n & \text{if } x \in [0, 1/n], \\ 0 & \text{if } x \in [1/n, 1]. \end{cases}$$

Then $\varphi_n(x) \to 0$ for all $x \in (0, 1]$ but $\|\varphi_n\|_1 = 1$.

In the next proposition we assume that μ is a finite measure, since we defined μ–uniform integrability only for finite measures μ.

Proposition 3.9. *Let (φ_n) be a sequence in $L^p(X, \mathcal{E}, \mu)$ pointwise convergent to a function $\varphi \in L^p(X, \mathcal{E}, \mu)$, with $(|\varphi_n|^p)$ μ–uniformly integrable. Then $\varphi_n \to \varphi$ in $L^p(X, \mathcal{E}, \mu)$.*

Proof. The functions $h_n := |\varphi_n - \varphi|^p$ are pointwise converging to 0 and, because of the inequality

$$h_n \leq 2^{p-1}(|\varphi_n|^p + |\varphi|^p),$$

they are also easily seen to be uniformly μ–integrable. Therefore, by applying Vitali Theorem 2.18 to h_n we obtain the conclusion. $\qquad\square$

3.4. The space $L^\infty(X, \mathcal{E}, \mu)$

Let $\varphi \colon X \to \overline{\mathbb{R}}$ be a $\mathcal{E}$–measurable function. We say that φ is μ–*essentially bounded* if there exists a real number $M > 0$ such that

$$\mu(\{|\varphi| > M\}) = 0.$$

If φ is μ–essentially bounded there exists a nonnegative number, denoted by $\|\varphi\|_\infty$, such that

$$\|\varphi\|_\infty = \min\left\{t \geq 0 \ : \ \mu(\{|\varphi| > t\}) = 0\right\}. \tag{3.8}$$

This easily follows from the fact that the function $t \to \mu(\{|\varphi| > t\})$ is right continuous (Proposition 2.6), so the infimum is attained.

Notice also that $\|\varphi\|_\infty$ is characterized by the property

$$\|\varphi\|_\infty \leq M \qquad \Longleftrightarrow \qquad |\varphi| \leq M \ \mu\text{–a.e. in } X. \tag{3.9}$$

We shall denote by $L^\infty(X, \mathcal{E}, \mu)$ the space of all equivalence classes of μ–essentially bounded functions with respect to the equivalence relation $\sim$ in (3.1), thus identifying functions that coincide μ–a.e. in X.

Several properties of the L^p spaces extend up to the case $p = \infty$: first of all $L^\infty(X, \mathcal{E}, \mu)$ is a real vector space and we have the Minkowski inequality

$$\|\varphi + \psi\|_\infty \leq \|\varphi\|_\infty + \|\psi\|_\infty. \tag{3.10}$$

Indeed, by (3.9) and the triangle inequality, $|\varphi(x) + \psi(x)| \leq \|\varphi\|_\infty + \|\psi\|_\infty$ μ–a.e. in X, therefore (3.8) provides (3.10). As a consequence, $L^\infty(X, \mathcal{E}, \mu)$ endowed with the norm $\| \cdot \|_\infty$, is a normed space.

The Hölder inequality takes the form

$$\int_X |\varphi\psi|\,d\mu \leq \|\varphi\|_\infty \int_X |\psi|\,d\mu. \tag{3.11}$$

Indeed, we have just to notice that $|\varphi(x)\psi(x)| \leq \|\varphi\|_\infty|\psi(x)|$ for μ–a.e. $x \in X$, and then integrate with respect to μ. This inequality can be still written as (3.6), provided we agree that $q = 1$ is the dual exponent of $p = \infty$ (and conversely).

For finite measures we can apply Hölder's inequality to obtain that the L^p spaces are nested; in particular L^∞ is the smaller one and L^1 is the larger one.

Remark 3.10 (Inclusions between L^p spaces). Assume that μ is finite. Then, if $1 \leq r \leq s \leq \infty$ we have

$$L^r(X, \mathscr{E}, \mu) \supset L^s(X, \mathscr{E}, \mu).$$

In fact, if $r < s$ and $\varphi \in L^s(X, \mathscr{E}, \mu)$ we have, in view of the Hölder inequality (with $p = s/r$ and $q = s/(s - r)$),

$$\int_X |\varphi(x)|^r\,d\mu(x) \leq \left(\int_X |\varphi(x)|^s\,d\mu(x)\right)^{r/s} \left(\int_X \mathbb{1}_X\,d\mu(x)\right)^{1-r/s},$$

and so

$$\|\varphi\|_r \leq (\mu(X))^{(s-r)/rs}\|\varphi\|_s. \tag{3.12}$$

By (3.12) we obtain that $p \mapsto \mu(X)^{-1/p}\|\varphi\|_p$ is nondecreasing for φ in the intersection of the spaces $L^p(X, \mathscr{E}, \mu)$, so that it has a limit as $p \to \infty$. Since $\mu(X)^{-1/p} \to 1$ as $p \to \infty$ we obtain that $\lim_{p\to\infty} \|\varphi\|_p$ exists, finite or infinite. The following proposition characterizes $L^\infty(X, \mathscr{E}, \mu)$ and the L^∞ norm in terms of this limit.

Proposition 3.11. *Assume that μ is finite and let φ be in the intersection*

$$\bigcap_{p<\infty} L^p(X, \mathscr{E}, \mu).$$

Then $\varphi \in L^\infty(X, \mathscr{E}, \mu)$ if and only if the limit $\lim_{p\to\infty} \|\varphi\|_p$ is finite. If this is the case, we have that $\|\varphi\|_\infty$ coincides with the value of the limit.

Proof. If $p \geq 1$ we have by the Markov inequality

$$\mu(\{|\varphi| \geq a\}) = \mu(\{|\varphi|^p \geq a^p\}) \leq a^{-p}\|\varphi\|_p^p.$$

Consequently, $\|\varphi\|_p \geq a\mu(\{|\varphi| \geq a\})^{1/p}$, which yields $\lim_p \|\varphi\|_p \geq a$ whenever $\mu(\{\varphi \geq a\}) > 0$. So, if the limit is finite, we have $\varphi \in$

$L^\infty(X, \mathscr{E}, \mu)$ and $\|\varphi\|_\infty \leq \lim_p \|\varphi\|_p$. The converse inequality follows directly from (3.11); the same inequality also proves that if the limit is not finite, then $\varphi \notin L^\infty(X, \mathscr{E}, \mu)$. $\qquad\qquad\square$

In the next remark we characterize the convergence in L^∞, proving also that $L^\infty(X, \mathscr{E}, \mu)$ is a Banach space: as a matter of fact, convergence in $L^\infty(X, \mathscr{E}, \mu)$ differs from the convergence in supremum norm only because a μ–negligible set is neglected.

Remark 3.12 ($L^\infty(X, \mathscr{E}, \mu)$ is a Banach space). Assume that $(\varphi_n) \subset L^\infty(X, \mathscr{E}, \mu)$ is a Cauchy sequence, and let us consider the μ–negligible set

$$\bigcup_{n,\,m=0}^{\infty} \{x \in X \ : \ |\varphi_n(x) - \varphi_m(x)| > \|\varphi_n - \varphi_m\|_\infty\}.$$

Then $\sup_{B^c} |\varphi_n - \varphi_m| \leq \|\varphi_n - \varphi_m\|_\infty$; as a consequence, the completeness of the space of bounded functions defined in B^c provides a bounded function $\varphi : B^c \to \mathbb{R}$ such that $\varphi_n \to \varphi$ uniformly in B^c. Extending φ in an arbitrary $\mathscr{E}$–measurable way (for instance with the 0 value) to the whole of X, we get $\varphi_n \to \varphi$ in $L^\infty(X, \mathscr{E}, \mu)$.
A similar argument proves that $\varphi_n \to \varphi$ in $L^\infty(X, \mathscr{E}, \mu)$ if and only if there exists a μ–negligible set $B \in \mathscr{E}$ satisfying $\varphi_n \to \varphi$ uniformly in B^c.

We know that $\left|\int_X \varphi \, d\mu\right|$ does not exceed $\int_X |\varphi| \, d\mu$. A nice and useful generalization of this fact is the so-called *Jensen inequality*.

Recall that, if $J \subset \mathbb{R}$ is an interval, a continuous function $g : J \to \mathbb{R}$ is said to be *convex* if

$$g\left(\frac{x+y}{2}\right) \leq \frac{g(x) + g(y)}{2} \qquad \forall x, y \in J. \qquad (3.13)$$

By several approximations (see Exercise 3.7) one can prove that a convex function f satisfies $g(tx+(1-t)y) \leq tg(x)+(1-t)g(y)$ for all $x, y \in J$ and $t \in [0, 1]$, and even that

$$g\left(\sum_{i=1}^{n} t_i x_i\right) \leq \sum_{i=1}^{n} t_i g(x_i) \quad \text{whenever } t_i \geq 0,\, x_i \in J \text{ and } \sum_{i=1}^{n} t_i = 1.$$

$$(3.14)$$

In the proof we use an elementary property of convex functions $g : \mathbb{R} \to \mathbb{R}$ satisfying $g(t) \to +\infty$ as $|t| \to +\infty$, namely the existence of a minimum point t_0; moreover, the function g is nondecreasing in $[t_0, +\infty)$ and nonincreasing in $(-\infty, t_0]$ (see Exercise 3.8).

Proposition 3.13 (Jensen). *Assume that μ is a probability measure. Let $g : \mathbb{R} \to \mathbb{R}$ be convex and bounded from below and let $\varphi \in L^1(X, \mathscr{E}, \mu)$. Then we have*

$$g\left(\int_X \varphi \, d\mu\right) \le \int_X g(\varphi) \, d\mu. \tag{3.15}$$

Proof. Let us first show (3.15) when φ is simple. Let

$$\varphi = \sum_{i=1}^n \alpha_i \mathbb{1}_{A_i},$$

where $n \ge 1$ is an integer, $\alpha_1, \ldots, \alpha_n \in \mathbb{R}$ and $A_1, \ldots, A_n$ are mutually disjoint sets in $\mathscr{E}$ whose union is X, so that

$$\sum_{i=1}^n \mu(A_i) = 1.$$

Then, from (3.14) we infer

$$g\left(\int_X \varphi \, d\mu\right) = g\left(\sum_{i=1}^n \alpha_i \mu(A_i)\right) \le \sum_{i=1}^n g(\alpha_i)\mu(A_i) = \int_X g(\varphi) \, d\mu.$$

In the general case, let us first assume that $g(t) \to +\infty$ as $|t| \to +\infty$. Then, by Exercise 3.8 we know that g has a minimum point t_0, and that g is nondecreasing in $[t_0, +\infty)$, and nonincreasing in $(-\infty, t_0]$. We can assume with no loss of generality (possibly replacing $g(t)$ by $g(t - t_0)$ and φ by $\varphi + t_0$) that g attains its minimum value at $t_0 = 0$, and that $\int_X g(\varphi) \, d\mu$ is finite. Furthermore, replacing g by $g - g(0)$, we can assume that the minimum value of g is 0.

Let $\varphi_n^\pm$ be nonnegative simple functions satisfying $\varphi_n^\pm \uparrow \varphi^\pm$; the simple functions $\varphi_n^+ - \varphi_n^-$ converge to $\varphi^+ - \varphi^- = \varphi$ in $L^1(X, \mathscr{E}, \mu)$. In addition, since g is monotone in $(-\infty, 0]$ and $[0, +\infty)$, the monotone convergence theorem gives

$$\int_X g(\varphi_n^+) \, d\mu \uparrow \int_X g(\varphi^+) \, d\mu, \qquad \int_X g(-\varphi_n^-) \, d\mu \uparrow \int g(-\varphi^-) \, d\mu,$$

so that (since $g(0) = 0$, $\varphi_n^+ \varphi_n^- = 0$ and $\varphi^+ \varphi^- = 0$) $\int_X g(\varphi_n^+ - \varphi_n^-) \, d\mu = \int_X g(\varphi_n^+) \, d\mu + \int_X g(-\varphi_n^-)$ converges to $\int_X g(\varphi^+) \, d\mu + \int_X g(-\varphi^-) = \int_X g(\varphi) \, d\mu$. Passing to the limit as $n \to \infty$ in Jensen's inequality for the simple functions $\varphi_n^+ - \varphi_n^-$

$$g\left(\int_X (\varphi_n^+ - \varphi_n^-) \, d\mu\right) \le \int_X g(\varphi_n^+ - \varphi_n^-) \, d\mu$$

we get (3.15).

Finally, the assumption that $g(t) \to +\infty$ as $t \to +\infty$ can be removed by considering the functions $g_\varepsilon(t) := g(t) + \varepsilon|t|$, which converge to $+\infty$ as $|t| \to \infty$, thanks to the fact that g is bounded from below: we obtain

$$g\left(\int_X \varphi \, d\mu\right) + \varepsilon \left|\int_X \varphi \, d\mu\right| \leq \int_X g(\varphi) \, d\mu + \varepsilon \int_X |\varphi| \, d\mu.$$

and Jensen's inequality follows by letting $\varepsilon \downarrow 0$. $\qquad\qquad\square$

An alternative proof of Jensen's inequality is based on another viewpoint, namely the representation of g as the supremum of a family $\{L_i\}_{i \in I}$ of affine functions. Since μ is a probability measure, for all $i \in I$ it is easy to check that $L_i(\int \varphi \, d\mu) = \int L_i(\varphi) \, d\mu$, so that

$$L_i\left(\int_X \varphi \, d\mu\right) \leq \int_X L(\varphi) \, d\mu \qquad \forall i \in I.$$

Taking the supremum in the right hand side we obtain Jensen's inequality. Both viewpoints are important in the theory of convex functions.

To be more precise, Jensen's inequality holds provided g is convex on an interval containing the image of φ. The next example is very important in Probability and Information theory.

Example 3.14 (Entropy functional). By applying Jensen's inequality with the convex function $g(z) = z \ln z$ in $[0, +\infty)$ we obtain

$$\int_X \varphi \ln \varphi \, d\mu \geq \int_X \varphi \, d\mu \ln\left(\int_X \varphi \, d\mu\right) \qquad (3.16)$$

for all $\varphi \in L^1(X, \mathscr{E}, \mu)$ nonnegative. If $\int_X \varphi \, d\mu = 1$ we obtain that $\int_X \varphi \ln \varphi \, d\mu \geq 0$ even though the function g has a variable sign (it attains the minimum value $-1/e$ at $z = 1/e$).

3.5. Dense subsets of $L^p(X, \mathscr{E}, \mu)$

Proposition 3.15. *For any $p \in [1, +\infty]$, the space of all simple μ–integrable functions is dense in $L^p(X, \mathscr{E}, \mu)$.*

Proof. Let $f \in L^p(X, \mathscr{E}, \mu)$ with $f \geq 0$. Then the conclusion follows from Proposition 2.12 (by Proposition 2.4 in the case $p = \infty$) and the dominated convergence theorem. In the general case we write f as $f^+ - f^-$ and approximate in L^p both parts by simple functions. $\qquad\square$

We consider now the special situation when X is a metric space, $\mathscr{E}$ is the σ–algebra of all Borel subsets of X and μ is any finite measure on $(X, \mathscr{E})$.

We denote by $C_b(X)$ the space of all continuous bounded functions on X. Clearly, $C_b(X) \subset L^p(X, \mathscr{E}, \mu)$ for all $p \in [1, +\infty]$.

Proposition 3.16. *For any $p \in [1, +\infty)$ and any finite measure μ, $C_b(X)$ is dense in $L^p(X, \mathcal{E}, \mu)$.*

Proof. Let $\mathscr{C}$ be the closure of $C_b(X)$ in $L^p(X, \mathcal{E}, \mu)$; obviously $\mathscr{C}$ is a vector space, as $C_b(X)$ is a vector space. In view of Proposition 3.15 it is enough to show that for any Borel set $I \in \mathscr{B}(X)$ there exists a sequence $(\varphi_n) \subset C_b(X)$ such that $\varphi_n \to \mathbb{1}_I$ in $L^p(X, \mathcal{E}, \mu)$.

Assume first that I is closed. Set

$$\varphi_n(x) = \begin{cases} 1 - n\, d(x, I) \text{ if } d(x, I) \leq \frac{1}{n} \\[2mm] 0 \text{ if } d(x, I) \geq \frac{1}{n}, \end{cases}$$

where

$$d(x, I) := \inf\{|x - y| : y \in I\}.$$

It is easy to see that φ_n are continuous, that $0 \leq \varphi_n \leq 1$ and that $\varphi_n(x) \to \mathbb{1}_I(x)$, hence the dominated convergence theorem implies that $\varphi_n \to \mathbb{1}_I$ in $L^p(X, \mathcal{E}, \mu)$.

Now, let

$$\mathscr{G} := \{I \in \mathscr{B}(X) : \mathbb{1}_I \in \mathscr{C}\}.$$

It is easy to see that $\mathscr{G}$ is a Dynkin system (which includes the π–system of closed sets), so that by the Dynkin theorem we have $\mathscr{G} = \mathscr{B}(X)$. $\square$

Remark 3.17. $C_b(X)$ (or more precisely, the equivalence classes of continuous bounded functions) is a closed subspace of $L^\infty(X, \mathcal{E}, \mu)$, and therefore it is not dense in general. Indeed, if $(\varphi_n) \subset C_b(X)$ is Cauchy in $L^\infty(X, \mathcal{E}, \mu)$, then it uniformly converges, up to a μ-negligible set B (just take in Remark 3.12 as B the union of the μ–negligible sets $\{|\varphi_n - \varphi_m| > \|\varphi_n - \varphi_m\|\}$). Therefore (φ_n) uniformly converges on B^c and on its closure K. Denoting by $\varphi \in C_b(K)$ its uniform limit, by Tietze's exension theorem we may extend φ to a function, that we still denote by φ, in $C_b(X)$. As $X \setminus K \subset B$ is μ–negligible, it follows that $\varphi_n \to \varphi$ in $L^\infty(X, \mathcal{E}, \mu)$.

Exercises

3.1 Assume that μ is σ–finite, but not finite. Provide examples showing that no inclusion holds between the spaces $L^p(X, \mathcal{E}, \mu)$ in general. Nevertheless, show that for any $\mathcal{E}$–measurable function $\varphi : X \to \mathbb{R}$ the set

$$\{p \in [1, \infty] : \varphi \in L^p(X, \mathcal{E}, \mu)\}$$

is an interval. *Hint:* consider for instance the Lebesgue measure on $\mathbb{R}$.

3.2 Let $1 \le p \le q < \infty$ and $f \in L^q(X, \mathscr{E}, \mu)$. Show that for any $\delta \in (0, 1)$ we can write $f = g + \tilde{f}$, with $g \in L^q(X, \mathscr{E}, \mu)$, $\tilde{f} \in L^p(X, \mathscr{E}, \mu)$ and $\|g\|_q \le \delta\|f\|_q$ (notice that if μ is finite we can take $g = 0$).

3.3 Let $p \in (1, \infty)$, $\varphi \in L^p$ and $\psi \in L^q$, with $q = p'$, be such that $\|\varphi\psi\|_1 = \|\varphi\|_p\|\psi\|_q$. Show that either $\psi = 0$ or there exists a constant $\lambda \in [0, +\infty)$ such that $|\varphi| = \lambda|\psi|^{q-1}$ μ–a.e. in X. *Hint:* first investigate the case of equality in Young's inequality.

3.4 Prove the following variant of Hölder's inequality, known as *Young*'s inequality: if $\varphi \in L^p$, $\psi \in L^q$ and $\frac{1}{p} + \frac{1}{q} = \frac{1}{r}$, with $r \ge 1$, we have that $\varphi\psi \in L^r$ and $\|\varphi\psi\|_r \le \|\varphi\|_p\|\psi\|_q$.

3.5 Let $(\varphi_n) \subset L^1(X, \mathscr{E}, \mu)$ be nonnegative and satisfying $\liminf_n \varphi_n \ge \varphi$ μ–a.e. in X. Show that

$$\int_X \varphi_n \, d\mu = \int_X \varphi \, d\mu = 1 \quad \Longrightarrow \quad \int_X |\varphi - \varphi_n| \, d\mu \to 0.$$

Hint: notice that the positive part and the negative part of $\varphi - \varphi_n$ have the same integral to obtain

$$\int_X |\varphi - \varphi_n| \, d\mu = 2 \int_X (\varphi - \varphi_n)^+ \, d\mu.$$

Then, apply the dominated convergence theorem.

3.6 Show that the following extension of Fatou's lemma: if $\varphi_n \ge -\psi_n$, with $\psi_n \in L^1(X)$ nonnegative, $\psi_n \to \psi$ in $L^1(X)$, then

$$\liminf_{n \to \infty} \int_X \varphi_n \, d\mu \ge \int_X \liminf_{n \to \infty} \varphi_n \, d\mu.$$

Hint: prove first the statement under the additional assumption that $\psi_n \to \psi$ μ–a.e. in X.

3.7 Show that (3.13) implies $g(tx + (1 - t)y) \le g(x) + (1 - t)g(y)$ for all $x, y \in J$ and $t \in [0, 1]$. Then, deduce from this property (3.14). *Hint:* it is useful to consider dyadic numbers $t = k/2^m$, with $k \le 2^m$ integer.

3.8 Let $g : \mathbb{R} \to \mathbb{R}$ be a convex function such that $g(z) \to +\infty$ as $|z| \to +\infty$. Show the existence of $z_0 \in \mathbb{R}$ where g attains its minimum value. Then, show that g is nondecreasing in $[z_0, +\infty)$ and nonincreasing in $(-\infty, z_0]$.

3.9 Let $(\varphi_n) \subset L^1(X, \mathscr{E}, \mu)$ be nonnegative functions. Show that the conditions

$$\liminf_{n \to \infty} \varphi_n \ge \varphi \quad \mu\text{–a.e. in } X, \qquad \limsup_{n \to \infty} \int_X \varphi_n \, d\mu \le \int_X \varphi \, d\mu < \infty$$

imply the convergence of φ_n to φ in $L^1(X, \mathscr{E}, \mu)$. *Hint:* use Exercise 3.5.

3.10 Let $\{\varphi_i\}_{i \in I}$ be a family of functions satisfying

$$\sup_{i \in I} \int_X \Phi(|\varphi_i|) \, d\mu = M < +\infty$$

and assume that $\Phi(c)/c$ is nondecreasing and tends to $+\infty$ as $c \to +\infty$. Show that $\{\varphi_i\}_{i \in I}$ is μ–uniformly integrable. *Hint:* use the inequalities

$$\int_A |\varphi_i|\, d\mu \leq \int_{A \cap \{|\varphi_i| \geq c\}} \frac{\Phi(\varphi_i)}{\Psi(c)}\, d\mu + \int_{A \cap \{|\varphi_i| < c\}} |\varphi_i|\, d\mu \leq \frac{M}{\Psi(c)} + c\mu(A),$$

with $\Psi(c) := \Phi(c)/c$, and then choose c sufficiently large, such that $M/\Psi(c) < \varepsilon/2$.

3.11 ★ Assuming that (X, d) is a metric space, $\mathscr{E} = \mathscr{B}(X)$ and μ is finite, prove *Lusin*'s theorem: for any $\varepsilon > 0$ and any $f \in L^1(X, \mathscr{E}, \mu)$, there exists a closed set $C \subset X$ such that $\mu(X \setminus C) < \varepsilon$ and $f|_C$ is continuous and bounded. *Hint:* use the density of $C_b(X)$ in L^1 and Egorov's theorem.

Chapter 4
Hilbert spaces

In this chapter we recall the basic facts regarding real vector spaces endowed with a scalar product. We introduce the concept of Hilbert space and show that, even for the infinite-dimensional ones, continuous linear functionals are induced by the scalar product. Moreover, we see that even in some classes of infinite dimensional spaces (the so-called separable ones) there exists a well-defined notion of basis (the so-called complete orthonormal systems), obtained replacing finite sums with converging series. Even though the presentation will be self-contained, we assume that the reader has already some familiarity with these concepts (basis, scalar product, representation of linear functionals) in finite-dimensional spaces.

4.1. Scalar products, pre-Hilbert and Hilbert spaces

A real *pre–Hilbert space* is a real vector space H endowed with a mapping

$$H \times H \to \mathbb{R}, \quad (x, y) \to \langle x, y \rangle,$$

called *scalar product*, such that:

(i) $\langle x, x \rangle \geq 0$ for all $x \in H$ and $\langle x, x \rangle = 0$ if and only if $x = 0$;
(ii) $\langle x, y \rangle = \langle y, x \rangle$ for all $x, y \in H$;
(iii) $\langle \alpha x + \beta y, z \rangle = \alpha \langle x, z \rangle + \beta \langle y, z \rangle$ for all $x, y, z \in H$ and $\alpha, \beta \in \mathbb{R}$.

In the following H represents a real pre–Hilbert space.

The scalar product allows us to introduce the concept of orthogonality. We say that two elements x and y of H are *orthogonal* if $\langle x, y \rangle = 0$.

We are going to prove that the function

$$\|x\| := \sqrt{\langle x, x \rangle}, \qquad x \in H$$

is a norm in H. For this we need the following *Cauchy–Schwartz* inequality.

Proposition 4.1. *For any x, $y \in H$ we have*

$$|\langle x, y \rangle| \leq \|x\| \, \|y\|. \tag{4.1}$$

In (4.1) equality holds if and only if x and y are linearly dependent.

Proof. Set

$$F(\lambda) = \|x + \lambda y\|^2 = \lambda^2 \|y\|^2 + 2\lambda \langle x, y \rangle + \|x\|^2, \qquad \lambda \in \mathbb{R}.$$

Since $F(\lambda) \geq 0$ for all $\lambda \in \mathbb{R}$ we have

$$|\langle x, y \rangle|^2 - \|x\|^2 \, \|y\|^2 \leq 0,$$

which yields (4.1).

If x and y are linearly dependent, it is clear that $|\langle x, y \rangle| = \|x\| \, \|y\|$. Assume conversely that $\langle x, y \rangle = \pm \|x\| \, \|y\|$ and that $y \neq 0$. Then we have $F(\lambda) = (\|x\| \pm \lambda \|y\|)^2$ so that, choosing $\lambda = \mp \|x\|/\|y\|$, we find $F(\lambda) = 0$. This implies $x + \lambda y = 0$, so that x and y are linearly dependent. $\qquad \square$

Now we can prove easily that $\| \cdot \|$ is a norm in H. In fact, it is clear that $\|\alpha x\| = |\alpha| \|x\|$ for all $\alpha \in \mathbb{R}$ and all $x \in H$. Moreover, taking into account (4.1), we have for all $x, y \in H$,

$$\|x + y\|^2 = \langle x + y, x + y \rangle = \|x\|^2 + \|y\|^2 + 2\langle x, y \rangle$$

$$\leq \|x\|^2 + \|y\|^2 + 2\|x\| \, \|y\| = (\|x\| + \|y\|)^2,$$

so that $\|x + y\| \leq \|x\| + \|y\|$.

Therefore a pre–Hilbert space H is a normed space and, in particular, a metric space. If H, endowed with the distance induced by the norm, is complete we say that H is a *Hilbert space*.

Example 4.2. (i). $\mathbb{R}^n$ is a Hilbert space with the canonical scalar product

$$\langle x, y \rangle := \sum_{k=1}^{n} x_k y_k,$$

inducing the Euclidean distance, where $x = (x_1, \ldots, x_n)$, $y = (y_1, \ldots, y_n) \in \mathbb{R}^n$.

(ii). Let $(X, \mathscr{E}, \mu)$ be a measure space. Then $L^2(X, \mathscr{E}, \mu)$, endowed with the scalar product

$$\langle \varphi, \psi \rangle := \int_X \varphi(x) \psi(x) \, d\mu(x) \qquad \varphi, \psi \in L^2(X, \mathscr{E}, \mu),$$

is a Hilbert space (completeness follows from Proposition 3.5).

(iii). Let ℓ^2 be the space of all sequences of real numbers $x = (x_k)$ such that $\sum_{k=0}^{\infty} x_k^2 < \infty$. ℓ^2 is a vector space with the usual operations,

$$a(x_k) = (ax_k) \quad a \in \mathbb{R}, \qquad (x_k) + (y_k) = (x_k + y_k), \qquad (x_k), (y_k) \in \ell^2.$$

The space ℓ^2, endowed with the scalar product

$$\langle x, y \rangle := \sum_{k=0}^{\infty} x_k y_k, \quad x = (x_k), \ y = (y_k) \in \ell^2$$

is a Hilbert space. This follows from (ii) taking $X = \mathbb{N}$, $\mathscr{E} = \mathscr{P}(X)$ and $\mu(\{x\}) = 1$ for all $x \in X$.

(iv). Let $X = C([0, 1])$ be the linear space of all real continuous functions on $[0, 1]$. X is a pre–Hilbert space with the scalar product

$$\langle f, g \rangle := \int_X f(t)g(t) \, dt.$$

However, X is not a Hilbert space: indeed, X is dense, but strictly contained, in $L^2(0, 1)$.

Finite-dimensional pre-Hilbert spaces H are always Hilbert spaces: indeed, if $\{v_1, \ldots, v_n\}$, with $n = \dim H$, is a basis of H, the Gram-Schmidt orthonormalization process (recalled in Exercise 4.3) provides an orthonormal basis $\{e_1, \ldots, e_n\}$ of H (i.e. $\|e_i\| = 1$ and e_i is orthogonal to e_j for $i \neq j$), and the map

$$x = \sum_{i=1}^{n} \langle x, e_i \rangle e_i \mapsto (\langle x, e_1 \rangle, \langle x, e_2 \rangle, \ldots, \langle x, e_n \rangle)$$

(mapping x to the Euclidean vector of its coordinates with respect to this basis) is easily seen to provide an isometry with $\mathbb{R}^n$: indeed,

$$\left\| \sum_{i=1}^{n} \langle x, e_i \rangle e_i \right\|^2 = \sum_{i, j=1}^{n} \langle x, e_i \rangle \langle x, e_j \rangle \langle e_i, e_j \rangle = \sum_{i=1}^{n} (\langle x, e_i \rangle)^2.$$

Thus, being $\mathbb{R}^n$ complete, H is complete.

4.2. The projection theorem

It is useful to notice that for any $x, y \in H$ the following *parallelogram identity* holds:

$$\|x + y\|^2 + \|x - y\|^2 = 2\|x\|^2 + 2\|y\|^2, \qquad x, y \in H. \tag{4.2}$$

One can show that identity (4.2) characterizes pre-Hilbert spaces among normed spaces, and Hilbert among Banach spaces, see Exercise 4.1.

Theorem 4.3 (Projection on closed subspaces). *Let H be a Hilbert space and let Y be a closed subspace of H. Then for any $x \in H$ there exists a unique $y \in Y$, called* projection of x on Y and denoted by $\pi_Y(x)$, *such that*

$$\|x - y\| = \min_{z \in Y} \|x - z\|.$$

Moreover, y is characterized by the property

$$\langle x - y, z \rangle = 0 \quad \text{for all } z \in Y. \tag{4.3}$$

Proof. Set $d := \inf_{z \in Y} \|x - z\|$ and choose $(y_n) \subset Y$ such that $\|x - y_n\| \downarrow d$. We are going to show that (y_n) is a Cauchy sequence.

For any $m, n \in \mathbb{N}$ we have, by the parallelogram identity (4.2),

$$\|(x - y_n) + (x - y_m)\|^2 + \|(x - y_n) - (x - y_m)\|^2 = 2\|x - y_n\|^2 + 2\|x - y_m\|^2.$$

Consequently

$$\|y_n - y_m\|^2 = 2\|x - y_n\|^2 + 2\|x - y_m\|^2 - 4\left\|x - \frac{y_n + y_m}{2}\right\|^2.$$

Taking into account that $(y_n + y_m)/2 \in Y$ we find

$$\|y_n - y_m\|^2 \le 2\|x - y_n\|^2 + 2\|x - y_m\|^2 - 4d^2,$$

so that $\|y_n - y_m\| \to 0$ as $n, m \to \infty$. Thus, (y_n) is a Cauchy sequence and, since the space is complete and Y is closed, it is convergent to an element $y \in Y$. Since $\|x - y_n\| \to \|x - y\|$ we find that $\|x - y\| = d$. Existence is thus proved. Uniqueness follows again by the parallelogram identity, that gives

$$\|y - y'\|^2 \le 2\|x - y\|^2 + 2\|x - y'\|^2 - 4\left\|x - \frac{y + y'}{2}\right\|^2$$
$$\le 2d^2 + 2d^2 - 4d^2 = 0$$

whenever y and y' are minimizers.

Let us prove (4.3). Define

$$F(\lambda) = \|x - y - \lambda z\|^2 = \lambda^2 \|z\|^2 - 2\lambda \langle x - y, z \rangle + \|x - y\|^2, \quad \lambda \in \mathbb{R}.$$

Since F attains a minimum at $\lambda = 0$, we have $F'(0) = \langle x - y, z \rangle = 0$, as claimed.

Conversely, if (4.3) holds for all $z \in Y$, we have

$$\|x - y - z\|^2 = \|z\|^2 + \|x - y\|^2 \ge \|x - y\|^2. \qquad \square$$

Remark 4.4 (Projection on convex closed sets). The previous proof works, with absolutely no modification, to show that for any convex closed set $K \subset H$ and any $x \in H$ there exists a unique solution $y = \pi_K(x)$ to the problem

$$\min_{z \in K} \|x - z\|.$$

In this case, however, $\pi_K(x)$ is not characterized by (4.3), but by a one-sided condition, namely $\langle x - \pi_K(x), z - \pi_K(x) \rangle \leq 0$ for all $z \in K$, see Exercise 4.2.

Corollary 4.5. *Let Y be a closed proper subspace of H. Then there exists $x_0 \in H \setminus \{0\}$ such that $\langle x_0, y \rangle = 0$ for all $y \in Y$.*

Proof. It is enough to choose an element z_0 in H which does not belong to Y and set $x_0 = z_0 - \pi_Y(z_0)$. $\qquad\qquad\square$

Fix an integer $n \geq 1$, a n-dimensional subspace $H_n \subset H$ and an orthonormal basis $\{e_1, \ldots, e_n\}$ of it. The following result characterizes the projection on H_n, giving the best approximation of an element x by a linear combination of $\{e_1, \ldots, e_n\}$.

Proposition 4.6. *The projection of an element $x \in H$ on H_n is given by*

$$\pi_{H_n}(x) = \sum_{k=1}^{n} \langle x, e_k \rangle e_k.$$

Proof. We have to show that for any $y_1, \ldots, y_n \in \mathbb{R}$ we have

$$\left\| x - \sum_{k=1}^{n} x_k e_k \right\|^2 \leq \left\| x - \sum_{k=1}^{n} y_k e_k \right\|^2, \qquad (4.4)$$

where $x_k = \langle x, e_k \rangle$. We have in fact

$$\left\| x - \sum_{k=1}^{n} y_k e_k \right\|^2 = \|x\|^2 + \sum_{k=1}^{n} y_k^2 - 2 \sum_{k=1}^{n} x_k y_k$$

$$= \|x\|^2 - \sum_{k=1}^{n} x_k^2 + \sum_{k=1}^{n} (x_k - y_k)^2.$$

This quantity is clearly minimal when $x_k = y_k$, and

$$\left\| x - \sum_{k=1}^{n} x_k e_k \right\|^2 = \|x\|^2 - \sum_{k=1}^{n} x_k^2. \qquad (4.5)$$

An alternative proof of the Proposition, based on the characterization (4.3) of $\pi_{H_n}(x)$, is proposed in Exercise 4.4. $\qquad\qquad\square$

4.3. Linear continuous functionals

A *linear functional F* on H is a mapping $F : H \to \mathbb{R}$ such that

$$F(\alpha x + \beta y) = \alpha F(x) + \beta F(y) \qquad \forall x, y \in H, \; \forall \alpha, \beta \in \mathbb{R}.$$

F is said to be *bounded* if there exists $K \geq 0$ such that

$$|F(x)| \leq K \|x\| \qquad \text{for all } x \in H.$$

Proposition 4.7. *A linear functional F is continuous if, and only if, it is bounded.*

Proof. It is obvious that if F is bounded then it is continuous (even Lipschitz continuous). Assume conversely that F is continuous and, by contradiction, that it is not bounded. Then for any $n \in \mathbb{N}$ there exists $x_n \in H$ such that $|F(x_n)| \geq n^2 \|x_n\|$. Setting $y_n = \frac{1}{n} x_n / \|x_n\|$ we have $\|y_n\| = \frac{1}{n} \to 0$, whereas $F(y_n) \geq n$, which is a contradiction. $\qquad\square$

The following basic *Riesz* theorem, gives an intrinsic representation formula of all linear continuous functionals.

Proposition 4.8. *Let F be a linear continuous functional on H. Then there exists a unique $x_0 \in H$ such that*

$$F(x) = \langle x, x_0 \rangle \qquad \forall x \in H. \tag{4.6}$$

Proof. Assume that $F \neq 0$ and let $Y = F^{-1}(0) = \operatorname{Ker} F$. Then $Y \neq H$ is closed (because F is continuous) and a vector space (because F is linear), so that by Corollary 4.5 there exists $z_0 \in H$ such that $F(z_0) = 1$ and

$$\langle z_0, z \rangle = 0 \quad \text{for all } z \in \operatorname{Ker} F.$$

On the other hand, for any $x \in H$ the element $z = x - F(x)z_0$ belongs to $\operatorname{Ker} F$ since $F(z) = F(x) - F(x)F(z_0) = 0$. Therefore

$$\langle z_0, x - F(x)z_0 \rangle = 0 \quad \text{for all } x \in H,$$

so that

$$\langle x, z_0 \rangle - F(x) \|z_0\|^2 = 0$$

and (4.6) follows setting $x_0 = z_0 / \|z_0\|^2$.

It remains to prove the uniqueness. Let $y_0 \in H$ be such that

$$F(x) = \langle x, x_0 \rangle = \langle x, y_0 \rangle, \quad x \in H.$$

Then, choosing $x = x_0 - y_0$ we find that $\|x_0 - y_0\|^2 = 0$, so that $x_0 = y_0$. $\qquad\square$

4.4. Bessel inequality, Parseval identity and orthonormal systems

Let us discuss the concept of *basis* in a Hilbert space H, assuming with no loss of generality that the dimension of H is not finite. We use Kronecker's notation δ_{hk}, equal to 1 for $h = k$ and equal to 0 if $h = k$.

Definition 4.9 (Orthonormal system). A sequence $(e_k)_{k \in \mathbb{N}} \subset H$ is called an *orthonormal system* if

$$\langle e_h, e_k \rangle = \delta_{h,k}, \quad h, k \in \mathbb{N}.$$

Proposition 4.10. *Let $(e_k)_{k \in \mathbb{N}}$ be an orthonormal system in H.*

(i) *For any $x \in H$ we have*

$$\sum_{k=0}^{\infty} |\langle x, e_k \rangle|^2 \le \|x\|^2. \tag{4.7}$$

(ii) *For any $x \in H$ the series $\sum_{k=0}^{\infty} \langle x, e_k \rangle e_k$ is convergent in H[1].*

(iii) *Equality holds in (4.7) holds if and only if*

$$x = \sum_{k=0}^{\infty} \langle x, e_k \rangle e_k. \tag{4.8}$$

Inequality (4.7) is called *Bessel inequality* and when the equality holds, *Parseval identity*.

Proof. (i) Let $n \in \mathbb{N}$. Then by (4.5) we have

$$\left\| x - \sum_{k=0}^{n} \langle x, e_k \rangle e_k \right\|^2 = \|x\|^2 - \sum_{k=0}^{n} |\langle x, e_k \rangle|^2, \tag{4.9}$$

so that (4.7) follows by the arbitrariness of n.

(ii) Let $n, \, p \in \mathbb{N}$ and set

$$s_n = \sum_{k=0}^{n} \langle x, e_k \rangle e_k.$$

---

[1] A series $\sum_{k=0}^{\infty} x_i$ of vectors in a Banach space E is said to be convergent if the sequence of the finite sums $\sum_{k=0}^{n} x_i$ is convergent in E

Then

$$\|s_{n+p} - s_n\|^2 = \left\| \sum_{k=n+1}^{n+p} \langle x, e_k \rangle e_k \right\|^2 = \sum_{k=n+1}^{n+p} |\langle x, e_k \rangle|^2.$$

Since the series $\sum_{k=0}^{\infty} |\langle x, e_k \rangle|^2$ is convergent by (i), the sequence (s_n) is Cauchy and the conclusion follows.

Passing to the limit as $n \to \infty$ in (4.9) we find

$$\left\| x - \sum_{k=0}^{\infty} \langle x, e_k \rangle e_k \right\|^2 = \|x\|^2 - \sum_{k=0}^{\infty} |\langle x, e_k \rangle|^2.$$

This proves statement (iii). $\qquad\qquad\qquad\qquad\qquad\qquad\qquad\square$

Definition 4.11 (Complete orthonormal system). An orthonormal system $(e_k)_{k \in \mathbb{N}}$ is called *complete* if

$$x = \sum_{k=0}^{\infty} \langle x, e_k \rangle e_k \qquad \forall x \in H.$$

Example 4.12. Let $H = \ell^2$ as in Example 4.2(iii). Then, it is easy to see that the system (e_k), where

$$e_k := (0, 0, \ldots, 0, 1, 0, 0, \ldots) \qquad \text{(with the digit 1 in the } k\text{-th position)}$$

is complete. Indeed, if $x = (x_k) \in \ell^2$ we have that $\langle x, e_i \rangle = x_i$ (the i-th component of the sequence x), so that

$$\left\| x - \sum_{k=0}^{n} \langle x, e_i \rangle e_i \right\|^2 = \sum_{k=n+1}^{\infty} x_k^2 \to 0.$$

We already noticed that $\mathbb{R}^n$ is the canonical model of n-dimensional Hilbert spaces H, because any choice of an orthonormal basis $\{v_1, \ldots, v_n\}$ of H induces the linear isometry

$$a \mapsto \sum_{i=1}^{n} a_i e_i$$

from $\mathbb{R}^n$ to H (which, as a consequence, preserves also the scalar product, by the parallelogram identity). For similar reasons, ℓ^2 is the canonical model of all spaces H having a complete orthonormal system $(e_k)_{k\in\mathbb{N}}$: in this case, the linear map from ℓ^2 to H given by

$$a \mapsto \sum_{i=0}^{\infty} a_i e_i$$

is an isometry, thanks to Parseval's identity.

Proposition 4.13 (Completeness criterion). *Let (e_n) be an orthonormal system. Then (e_n) is complete if and only if the vector space E spanned by (e_n) is dense in H.*

Proof. If (e_n) is complete we have that any $x \in H$ is the limit of the finite sums $\sum_1^N \langle x, e_i \rangle e_i$, which all belong to E, therefore E is dense. Conversely, if E is dense, for any $x \in H$ and any $\varepsilon > 0$ we can find a vector $z = \sum_{i=1}^n a_i e_i$ with $\|z - x\| < \varepsilon$. By applying Proposition 4.6 twice (first to the vector space spanned by $\{e_1, \dots, e_m\}$, and then to the vector space spanned by $\{e_1, \dots, e_n\}$) we get

$$\left\| x - \sum_{i=1}^m \langle x, e_i \rangle e_i \right\| \le \left\| x - \sum_{i=1}^n \langle x, e_i \rangle e_i \right\| \le \left\| x - \sum_{i=1}^n a_i e_i \right\| < \varepsilon$$

for $m \ge n$. Since ε is arbitrary this proves that the sum of the series is equal to x. $\qquad\square$

The following proposition provides a necessary and sufficient condition for the existence of a complete orthonormal system. We recall that a metric space (X, d) is said to be *separable* if there exists a countable dense subset $D \subset X$.

Theorem 4.14. *A Hilbert space H admits a complete orthonormal system $(e_k)_{k\in\mathbb{N}}$ if and only if H, as a metric space, is separable.*

Proof. If H admits a complete orthonormal system $(e_k)_{k\in\mathbb{N}}$ then H is separable, because the collection $\mathscr{D}$ of finite sums with rational coefficients of the vectors e_k provides a countable dense subset (indeed, the closure of $\mathscr{D}$ contains the finite linear combinations of the vectors e_k and then the whole space).

Conversely, assume that H is separable and let (v_n) be a dense sequence. We define $e_0 = v_0$, $e_1 = v_{k_1}$ where k_1 is the first $k > k_0 = 0$ such that v_k is linearly independent from v_0, $e_2 = v_{k_2}$ where k_2 is the first $k > k_1$ such that v_k is linearly independent from $\{e_0, e_1\}$, and so on. In this way we have built a sequence (e_i) of linearly independent vectors generating the same vector space generated by (v_n). Let S be this vector space, and let us represent it as $\cup_n S_n$, where S_n is the vector space generated by $\{e_0, \ldots, e_n\}$. Notice that S is dense, as all v_n belong to S.

By applying the Gram-Schmidt process to e_i, an operation that does not change the vector spaces S_n generated by the vectors $e_0, \ldots, e_n$, we can also assume that (e_i) is an orthonormal system. Then, Proposition 4.13 gives that (e_i) is complete. $\qquad\square$

4.5. Hilbert spaces on $\mathbb{C}$

In this section we illustrate briefly how the concepts introduced so far extend to *complex* vector spaces H. A *pre–Hilbert space* is a complex vector space H endowed with a mapping

$$H \times H \to \mathbb{C}, \quad (x, y) \to \langle x, y \rangle,$$

called *scalar product*, such that:

(i) $\langle x, x \rangle \geq 0$ for all $x \in H$ and $\langle x, x \rangle = 0$ if and only if $x = 0$;
(ii) $\langle x, y \rangle = \overline{\langle y, x \rangle}$ for all $x, y \in H$;
(iii) $\langle \alpha x + \beta y, z \rangle = \alpha \langle x, z \rangle + \beta \langle y, z \rangle$ for all $x, y, z \in H$ and $\alpha, \beta \in \mathbb{C}$.

It turns out that $\|x\| := \sqrt{\langle x, x \rangle}$ is still a norm, because the Cauchy-Schwarz inequality still holds. Hence, we can define Hilbert spaces as those spaces for which the norm induces a complete distance.

The canonical model of n-dimensional Hilbert space is $\mathbb{C}^n$. Given a measure space $(X, \mathscr{F}, \mu)$, a basic example of Hilbert space is the space of $\mathscr{F}$-measurable and square integrable functions $f : X \to \mathbb{C}$. In this context $\mathscr{F}$-measurable means that both the real and the imaginary part of f are $\mathscr{F}$-measurable. In this space one can define the scalar product

$$\langle f, g \rangle := \int_X f(x)\overline{g(x)}\, d\mu(x)$$

and prove that it induces an Hilbert space structure. The space $\ell_2(\mathbb{C})$ of complex-values sequences (z_n) with $(|z_n|) \in \ell_2(\mathbb{R})$ is a particular case.

The norm still satisfies the parallelogram identity, so that we can still prove the existence of orthogonal projections on closed subspaces and its characterization in terms of

$$\mathrm{Re}\big(\langle x - \pi_Y(x), z \rangle\big) = 0 \qquad \forall z \in Y.$$

Analogously, in Remark 4.4, one has to replace the scalar product by its real part.

Riesz representation theorem still holds (now for continuous and $\mathbb{C}$-linear functionals) and the concepts of orthonormal system and complete orthonormal system make sense. We have Bessel's inequality for orthonormal systems and Parseval's identity for complete orthonormal systems. Finally, $\ell_2(\mathbb{C})$ is the canonical model of all separable Hilbert spaces; as in the real case the correspondence is induced by the choice of a complete orthonormal system, which provides coordinates of a vector.

We conclude this chapter providing a natural example, considered in the literature, of non-separable Hilbert space.

Example 4.15 (Quasi-periodic functions). We define the space $AP(\mathbb{R})$ of *almost periodic* functions as the closure, with respect to uniform convergence in $\mathbb{R}$, of the vector space generated by complex-valued periodic functions (of arbitrary period). This space has been extensively studied by Bochner and Bohr. It is easy to show that the space of almost periodic functions is not only a vector space (it is a subspace of $C(\mathbb{R}, \mathbb{C})$), but also an algebra, *i.e.* $fg \in AP(\mathbb{R})$ whenever $f, g \in AP(\mathbb{R})$.

If f is almost periodic one can also show (by approximation, taking into account that this property is linear with respect to f and holds for periodic functions) that there exists the limit

$$M(f) := \lim_{T \to +\infty} \frac{1}{2T} \int_{-T}^{T} f(x+t)\, dt.$$

In addition, it is easily seen that the limit is independent of x.

The space $AP(\mathbb{R})$ of all almost periodic functions is a pre-Hilbert space when endowed with the following inner product

$$\langle f, g \rangle_{AP} := M(f\bar{g}) \qquad f, g \in AP(\mathbb{R}).$$

For any $\lambda \in \mathbb{R}$ define

$$e_\lambda(t) = e^{i\lambda t}, \qquad t \in \mathbb{R}.$$

Then $e_\lambda \in AP(\mathbb{R})$, $\langle e_\lambda, e_\lambda \rangle_{AP} = 1$ and

$$\langle e_\lambda, e_\nu \rangle_{AP} = \lim_{T \to +\infty} \frac{e^{iT(\lambda-\nu)} - e^{-iT(\lambda-\nu)}}{Ti(\lambda - \nu)} = 0 \qquad \text{whenever } \lambda \neq \nu,$$

so that $(e_\lambda)_{\lambda \in \mathbb{R}}$ is an orthonormal system in $AP(\mathbb{R})$ having the cardinality of continuum. One can also characterize the (abstract) Hilbert completion of $AP(\mathbb{R})$ (the so-called Bohr almost periodic functions) and prove that the system $\{e_\lambda\}_{\lambda \in \mathbb{R}}$ is complete. For more details see *e.g.* [4].

Exercises

4.1 Let $(X, \| \cdot \|)$ be a normed space, and assume that the norm satisfies the parallelogram identity (4.2). Set

$$\langle x, y \rangle := \frac{1}{4} \| x + y \|^2 - \frac{1}{4} \| x - y \|^2, \quad x, y \in X.$$

Show that $\langle \cdot, \cdot \rangle$ is a scalar product whose induced norm is $\| \cdot \|$. Use this identity to show that any linear isometry between pre-Hilbert spaces preserves also the scalar product.

4.2 Show that, in the situation considered in Remark 4.4, $\pi_K(x)$ is characterized by the property

$$\langle x - \pi_K(x), z - \pi_K(x) \rangle \leq 0 \qquad \forall z \in K.$$

4.3 Let H be a finite dimensional pre-Hilbert space and let $\{v_1, \ldots, v_n\}$, with $n = \dim H$, be a basis of it. Define

$$f_1 = v_1, \quad f_2 = v_2 - \frac{\langle v_2, f_1 \rangle}{\langle f_1, f_1 \rangle} f_1, \quad f_3 = v_3 - \frac{\langle v_3, f_1 \rangle}{\langle f_1, f_1 \rangle} f_1 - \frac{\langle v_3, f_2 \rangle}{\langle f_2, f_2 \rangle} f_2, \ldots\ldots$$

Show that $e_i = f_i / \| f_i \|$ is an orthonormal system in H (notice that $v_k - f_k$ is the projection of v_k on the vector space generated by $\{v_1, \ldots, v_{k-1}\}$).

4.4 Let H be a Hilbert space, and let X be an infinite-dimensional separable subspace. Show that

$$\pi_X(x) = \sum_{k=0}^{\infty} \langle x, e_k \rangle e_k \qquad \forall x \in H,$$

where (e_k) is any complete orthonormal system of X. *Hint:* show that the vector $x - \sum_k \langle x, e_k \rangle e_k$ is orthogonal to all vectors of X.

4.5 Let X be the space of functions $f : [0, 1] \to \mathbb{R}$ such that $f(x) \neq 0$ for at most countably many x, and $\sum_x f^2(x) < +\infty$. Show that X, endowed with the scalar product

$$\langle f, g \rangle := \sum_{x \in [0,1]} f(x) g(x),$$

is a non-separable Hilbert space.

4.6 Let $(e_k)_{k \in \mathbb{N}}$ be a complete orthonormal system of H. Show that, for any $x, y \in H$ we have

$$\sum_{k=0}^{\infty} \langle x, e_k \rangle \langle y, e_k \rangle = \langle x, y \rangle. \tag{4.10}$$

4.7 $\star$ Show that for any Hilbert space H there exists a family (not necessarily finite or countable) of vectors $\{e_i\}_{i \in I}$ such that:

(i) $\langle e_i, e_j \rangle$ is equal to 1 if $i = j$, and to 0 otherwise;
(ii) for any vector $x \in H$ there exists a countable set $J \subset I$ with

$$x = \sum_{i \in J} \langle x, e_i \rangle e_i.$$

Hint: use Zorn's lemma.

Chapter 5
Fourier series

In this chapter we study the problem of representing a given T-periodic function as a superposition, for a suitable choice of the coefficients, of more "elementary" ones. This problem was first studied by J. Fourier in the case when the elementary functions are the trigonometric ones (nowadays we know that many different choices are indeed possible). Thanks to the theory of L^2 spaces and of Hilbert spaces developed in the previous chapters, the problem can be formalized by looking for complete orthonormal systems in L^2 made by trigonometric functions.

We shall mostly be concerned with the case of 2π-periodic functions, but a simple change of scale (see Remark 5.1) easily provides the translation of the results to arbitrary periods.

We are concerned with the measure space $\big((-\pi, \pi), \mathscr{B}((-\pi, \pi)), \lambda\big)$, where λ is the Lebesgue measure. As usual, we shall write for brevity $L^2(-\pi, \pi)$. We shall denote by $\langle \cdot, \cdot \rangle$ the canonical scalar product given by

$$\langle f, g \rangle := \int_{(-\pi,\pi)} f(x)g(x)\, d\lambda = \int_{-\pi}^{\pi} f(x)g(x)\, dx, \quad f, g \in L^2(-\pi, \pi).$$

Let us consider, as a family of elementary functions, the *trigonometric system*, given by:

$$\frac{1}{\sqrt{2\pi}}; \quad \frac{1}{\sqrt{\pi}}\cos kx, \ k \in \mathbb{N}, \ k \geq 1; \quad \frac{1}{\sqrt{\pi}}\sin kx, \ k \in \mathbb{N}, \ k \geq 1.$$

$$(5.1)$$

It is easy to check with integration by parts that this is an orthonormal system in $L^2(-\pi, \pi)$, see Exercise 5.1. Thus, in view of Proposition 4.10, the series of functions

$$S(x) = \frac{1}{2}a_0 + \sum_{k=1}^{\infty}(a_k \cos kx + b_k \sin kx), \tag{5.2}$$

is convergent in $L^2(-\pi, \pi)$ for any $f \in L^2(-\pi, \pi)$, where

$$a_k := \frac{1}{\pi} \int_{-\pi}^{\pi} f(y) \cos ky\, dy, \quad k \in \mathbb{N},$$

and

$$b_k := \frac{1}{\pi} \int_{-\pi}^{\pi} f(y) \sin ky\, dy, \quad k \in \mathbb{N}, \ k \geq 1.$$

Notice that $a_0/2$ is the mean value of f on $(-\pi, \pi)$, in agreement with the fact that all terms in the series (5.2) have mean value 0 on $(-\pi, \pi)$. To recognize (5.2) in terms of scalar products, we see that the term $a_0/2$ corresponds to

$$\left\langle f, \frac{1}{\sqrt{2\pi}} \right\rangle \frac{1}{\sqrt{2\pi}}$$

and the terms $a_k \cos kx$, $b_k \sin kx$ for $k \geq 1$, correspond respectively to

$$\left\langle f, \frac{1}{\sqrt{\pi}} \cos kx \right\rangle \frac{1}{\sqrt{\pi}} \cos kx, \qquad \left\langle f, \frac{1}{\sqrt{\pi}} \sin kx \right\rangle \frac{1}{\sqrt{\pi}} \sin kx.$$

Formula (5.2) is called the *trigonometric Fourier series* of f.

The Bessel inequality (4.7) reads, in this context, as follows:

$$\frac{1}{\pi} \int_{-\pi}^{\pi} |f(x)|^2 \, dx \geq \frac{1}{2} a_0^2 + \sum_{k=1}^{\infty} (a_k^2 + b_k^2). \tag{5.3}$$

Indeed, it is easily seen that $a_0^2 \pi/2 = (\langle f, 1/\sqrt{2\pi} \rangle)^2$ and, for $k \geq 1$,

$$a_k^2 \pi = \left(\left\langle f, \frac{1}{\sqrt{\pi}} \cos kx \right\rangle \right)^2, \qquad b_k^2 \pi = \left(\left\langle f, \frac{1}{\sqrt{\pi}} \sin kx \right\rangle \right)^2.$$

First, we shall find sufficient conditions on f ensuring the pointwise convergence of the series $S(x)$ to $f(x)$ in $(-\pi, \pi)$. Then, we shall show that the trigonometric system is complete, so that the inequality above is actually an equality. As shown in Exercise 5.4 and Exercise 5.5, the trigonometric system, the trigonometric series and the form of the coefficients become much more nice and symmetric in the complex-valued Hilbert space $L^2\big((-\pi, \pi); \mathbb{C}\big)$:

$$f(x) = \sum_{n \in \mathbb{Z}} a_n e^{inx} \qquad \text{where } a_n := \frac{1}{2\pi} \int_{-\pi}^{\pi} f(x) e^{-inx} \, dx.$$

Remark 5.1 (*$2T$-periodic functions*). If $f \in L^2(-T, T)$ we can write instead

$$f(x) = \frac{a_0}{2} + \sum_{k=1}^{\infty} a_k \cos \frac{\pi}{T} kx + b_k \sin \frac{\pi}{T} kx$$

with

$$a_k := \begin{cases} \dfrac{1}{T} \displaystyle\int_{-T}^{T} f(x)\, dx & \text{if } k = 0; \\[2ex] \dfrac{1}{T} \displaystyle\int_{-T}^{T} f(x) \cos \dfrac{\pi}{T} kx \, dx & \text{if } k > 0, \end{cases}$$

$$b_k := \frac{1}{T} \int_{-T}^{T} f(x) \sin \frac{\pi}{T} kx \, dx.$$

5.1. Pointwise convergence of the Fourier series

For any integer $N \geq 1$ we consider the partial sum

$$S_N(x) := \frac{1}{2} a_0 + \sum_{k=1}^{N} (a_k \cos kx + b_k \sin kx), \qquad x \in [-\pi, \pi).$$

Since the functions $\cos kx$ and $\sin kx$ are 2π–periodic, it is natural to extend f to the whole of $\mathbb{R}$ as a 2π–periodic function $\widetilde{f}$, setting

$$\widetilde{f}(x + 2\pi n) = f(x), \quad x \in [-\pi, \pi), \ n = \pm 1, \pm 2, \ldots . \tag{5.4}$$

We shall denote in the sequel by $H_{l,r}(z)$ the "Heaviside" function

$$H_{l,r}(z) := \begin{cases} l & \text{if } z \leq 0; \\ r & \text{if } z > 0. \end{cases}$$

Lemma 5.2. *For any integer $N \geq 1$ and $x,\, l,\, r \in \mathbb{R}$ we have*

$$S_N(x) - \frac{l+r}{2} = \frac{1}{2\pi} \int_{-\pi}^{\pi} \frac{\widetilde{f}(x + \tau) - H_{l,r}(\tau)}{\sin(\tau/2)} \sin\left[\left(N + \frac{1}{2}\right)\tau\right] d\tau. \tag{5.5}$$

Proof. Write

$$S_N(x) = \frac{1}{2} a_0 + \sum_{k=1}^{N} (a_k \cos kx + b_k \sin kx)$$

$$= \frac{1}{\pi} \int_{-\pi}^{\pi} f(y) \left[\frac{1}{2} + \sum_{k=1}^{N} (\cos kx \cos ky + \sin kx \sin ky) \right] dy$$

$$= \frac{1}{\pi} \int_{-\pi}^{\pi} f(y) \left[\frac{1}{2} + \sum_{k=1}^{N} \cos k(x-y) \right] dy.$$

To evaluate the sum, we notice that for any $z \in \mathbb{R}$

$$\left[\tfrac{1}{2} + \sum_{k=1}^{N} \cos kz \right] \sin\left(\tfrac{1}{2}z\right)$$

$$= \frac{1}{2} \left[\sin\left(\tfrac{1}{2}z\right) + \sum_{k=1}^{N} \left(\sin\left[(k + \tfrac{1}{2})z \right] - \sin\left[(k - \tfrac{1}{2})z \right] \right) \right]$$

$$= \frac{1}{2} \sin\left[(N + \tfrac{1}{2})z \right].$$

Therefore

$$\frac{1}{2} + \sum_{k=1}^{N} \cos kz = \frac{1}{2} \frac{\sin\left[\left(N + \tfrac{1}{2} \right) z \right]}{\sin\left(\tfrac{1}{2}z \right)} \tag{5.6}$$

and so,

$$S_N(x) = \frac{1}{2\pi} \int_{-\pi}^{\pi} f(y) \frac{\sin\left[\left(N + \tfrac{1}{2} \right) (x-y) \right]}{\sin\left(\tfrac{1}{2}(x-y) \right)} dy. \tag{5.7}$$

Now, setting $\tau = y - x$ we get

$$S_N(x) = \frac{1}{2\pi} \int_{-\pi-x}^{\pi-x} \widetilde{f}(x+\tau) \frac{\sin\left[\left(N + \tfrac{1}{2} \right) \tau \right]}{\sin\left(\tfrac{1}{2}\tau \right)} d\tau$$

$$= \frac{1}{2\pi} \int_{-\pi}^{\pi} \widetilde{f}(x+\tau) \frac{\sin\left[\left(N + \tfrac{1}{2} \right) \tau \right]}{\sin\left(\tfrac{1}{2}\tau \right)} d\tau$$

since the function under the integral is 2π–periodic. Now, integrating (5.6) over $[-\pi, \pi]$ yields

$$1 = \frac{1}{2\pi} \int_{-\pi}^{\pi} \frac{\sin\left[\left(N + \tfrac{1}{2} \right) \tau \right]}{\sin\left(\tfrac{1}{2}\tau \right)} d\tau,$$

so that

$$\frac{1}{\pi} \int_0^\pi \frac{\sin\left[\left(N + \frac{1}{2}\right)\tau\right]}{\sin\left(\frac{1}{2}\tau\right)}\, d\tau = 1 = \frac{1}{\pi} \int_{-\pi}^0 \frac{\sin\left[\left(N + \frac{1}{2}\right)\tau\right]}{\sin\left(\frac{1}{2}\tau\right)}\, d\tau.$$

If we multiply both sides by l and r, and subtract the resulting identities from (5.7), (5.5) follows. $\qquad\qquad\square$

Proposition 5.3 (Dini's test). *Let x, l, $r \in \mathbb{R}$ be such that*

$$\int_{-\pi}^\pi \frac{|\widetilde{f}(x + \tau) - H_{l,r}(\tau)|}{|\sin(\tau/2)|}\, d\tau < \infty. \tag{5.8}$$

Then the Fourier series of f converges to $(l + r)/2$ at x.

Dini's test shows a remarkable property of the Fourier series: while the specific value of the coefficients a_k and b_k depends on the behaviour of f on the *whole* interval $(-\pi, \pi)$, and the same holds for the Fourier series, the character of the series (convergent or not) at a given point x depends *only* on the behaviour of f in the neighbourhood of x: indeed, it is this behaviour that influences the integrability of $(\widetilde{f}(x + \tau) - H_{l,r}(\tau))/\sin(\tau/2)$ (the only singularity being at $\tau = 0$).

In the next example we provide sufficient conditions for the convergence of the Fourier series.

Example 5.4. Assume that $f : [-\pi, \pi] \to \mathbb{R}$ is L-Lipschitz continuous, *i.e.*

$$|f(x) - f(y)| \le L|x - y| \qquad \forall\, x,\, y \in [-\pi, \pi]$$

for some $L \ge 0$. Then Dini's test is fulfilled at any $x \in \mathbb{R} \setminus \mathbb{Z}\pi$ choosing $l = r = \widetilde{f}(x)$, and at any $x \in \mathbb{Z}\pi$ choosing $l = \widetilde{f}(x_-)$ and $r = \widetilde{f}(x_+)$ [1]. Indeed, with these choices of l and r, the quotient

$$\frac{\widetilde{f}(x + \tau) - H_{l,r}(\tau)}{\sin(\tau/2)}$$

is bounded in a neighbourhood of 0.
The same conclusions hold when f is α–Hölder continuous for some $\alpha \in (0, 1]$, *i.e.*

$$|f(x) - f(y)| \le L|x - y|^\alpha, \qquad \forall\, x,\, y \in [-\pi, \pi]$$

for some $L \ge 0$: in this case the quotient is bounded from above, near 0, by the function $L|\tau|^\alpha/|\sin(\tau/2)| \sim 2L|\tau|^{\alpha-1}$ which is integrable.

---

[1] here we denote by $g(x_-)$, $g(x_+)$ the left and right limits of g at x

More generally, the argument of the previous example can be used to show that the Fourier series is pointwise convergent for piecewise C^1 functions f: at continuity points x the series converges to $f(x)$, and at (jump) discontinuity points x it converges to $(f(x_-) + f(x_+))/2$. However, the mere continuity of f *is not* sufficient to ensure pointwise convergence of the Fourier series.

In order to prove Proposition 5.3, we need the following *Riemann–Lebesgue* lemma, a tool interesting in itself.

Lemma 5.5. *Let (e_k) be an orthonormal system in $L^2(-\pi, \pi)$. Assume that there exists $M > 0$ such that $\|e_k\|_\infty \leq M$ for all $k \in \mathbb{N}$. Then for any $f \in L^1(-\pi, \pi)$ we have*

$$\lim_{k \to \infty} \int_{-\pi}^{\pi} f(x) e_k(x)\, dx = 0. \tag{5.9}$$

Proof. Notice first that if $f \in L^2(-\pi, \pi)$ the conclusion of the lemma is trivial. We have in fact in this case

$$\int_{-\pi}^{\pi} f(x) e_k(x)\, dx = \langle f, e_k \rangle$$

and, since by Bessel's inequality the series $\sum_1^\infty |\langle f, e_k \rangle|^2$ is convergent, we have $\lim_k \langle f, e_k \rangle = 0$.

Let us now consider the general case. We know that bounded continuous functions are dense in $L^1(-\pi, \pi)$, hence for any $\varepsilon > 0$ we can find $g \in C_b(-\pi, \pi)$ such that $\|f - g\|_1 < \varepsilon$. As a consequence

$$|\langle f, e_k \rangle| = |\langle f - g, e_k \rangle| + |\langle g, e_k \rangle| \leq M\varepsilon + |\langle g, e_k \rangle|$$

and letting $k \to \infty$ we obtain $\limsup_k |\langle f, e_k \rangle| \leq M\varepsilon$. Since ε is arbitrary the proof is achieved. $\square$

Proof of Proposition 5.3. Set

$$g(\tau) := \frac{\widetilde{f}(x + \tau) - H_{l,r}(\tau)}{\sin(\tau/2)} \in L^1(-\pi, \pi). \tag{5.10}$$

Then, writing

$$\sin[(N + \tfrac{1}{2})t] = \sin Nt \cos \tfrac{1}{2}t + \cos Nt \sin \tfrac{1}{2}t$$

and applying the Riemann–Lebesgue lemma to $g \cos t/2$ (with $e_N = \sin Nt$) and to $g \sin(t/2)$ (with $e_N = \cos Nt$) we obtain from (5.5) that $S_N(x)$ converge to $(l + r)/2$. $\square$

5.2. Completeness of the trigonometric system

Proposition 5.6. *The trigonometric system* (5.1) *is complete. In particular equality holds in* (5.3) *and*

$$\lim_{N \to \infty} \int_{-\pi}^{\pi} |f(x) - S_N f(x)|^2 \, dx = 0 \qquad \forall f \in L^2(-\pi, \pi). \qquad (5.11)$$

Proof. We show that the vector space E generated by the trigonometric system is dense in $L^2(-\pi, \pi)$. Let H' be the closure, in the $L^2(-\pi, \pi)$ norm, of E, that is easily seen to be still a vector space as well. We will prove in a series of steps that H' contains larger and larger classes of functions.

Let $f : [-\pi, \pi] \to [0, +\infty)$ be a Lipschitz function, and let us prove that it belongs to H'. Indeed, we know from Example 5.4 that $S_N \to f$ pointwise in $(-\pi, \pi)$. On the other hand, we already know from Proposition 4.10(ii) that the Fourier series is convergent in $L^2(-\pi, \pi)$ to some function g (which is indeed, by Exercise 4.4, the orthogonal projection of f on H'), therefore a subsequence $(S_{N(k)})$ is converging λ-almost everywhere to g. It follows that $g = f$ and $S_N \to f$ in $L^2(-\pi, \pi)$.

If now $g : [-\pi, \pi] \to [0, +\infty)$ is continuous, we know that g can be monotonically approximated by the Lipschitz functions

$$g_\lambda(x) := \min_{y \in [-\pi, \pi]} \big(g(y) + \lambda |x - y|\big), \qquad x \in [-\pi, \pi]$$

(see Exercise 2.11), that converge to g also in $L^2(-\pi, \pi)$ by the dominated convergence theorem. As a consequence also g belongs to H'. Since H' is invariant by addition of constants, we proved that all continuous functions in $[-\pi, \pi]$ belong to H'. We conclude using the density of this class of functions in $L^2(-\pi, \pi)$. $\qquad \square$

Remark 5.7. Let $f \in L^2(-\pi, \pi)$. Then, the Parseval identity reads as follows

$$\frac{1}{\pi} \int_{-\pi}^{\pi} |f(x)|^2 \, dx = \frac{1}{2} a_0^2 + \sum_{k=1}^{\infty} (a_k^2 + b_k^2). \qquad (5.12)$$

For instance, taking $f(x) = x$ one finds the following nice relation between π and the harmonic series with exponent 2:

$$\sum_{k=1}^{\infty} \frac{1}{k^2} = \frac{\pi^2}{6}.$$

Notice that (5.11) provides, for any $f \in L^2(-\pi, \pi)$, the existence of a subsequence $N(k)$ such that $S_{N(k)} f(x) \to f(x)$ for $\mathscr{L}^1$–a.e. $x \in$

$(-\pi, \pi)$. Is it true that the *whole* sequence $S_N f$ converges a.e. to f? This problem, surprisingly difficult, has been solved by L.Carleson only in 1966, see [1].

Finally, we notice that there exist other important examples of complete orthonormal systems, besides the trigonometric one. Some of them are illustrated in the exercises.

5.3. Uniform convergence of the Fourier series

We conclude by studying the uniform convergence of the Fourier series. We recall that a series $\sum_0^\infty x_n$ in a Banach space E is said to be totally convergent if the numerical series $\sum_0^\infty \|x_n\|$ is convergent. Using the completeness of E it is not difficult to check (see Exercise 5.2) that any totally convergent series is convergent (as we have seen in the previous chapter, this means that the finite sums $\sum_0^N x_n$ converge in E to a vector, denoted by $\sum_0^\infty x_n$).

Now we show that the Fourier series of C^1 functions f with $f(-\pi) = f(\pi)$ are uniformly convergent: the proof highlights two important principles, whose validity extend to higher order derivatives (see Exercise 5.9) and to Fourier transforms: first, the Fourier coefficients of the derivative of a function are linked to the Fourier coefficients of the function; second, higher regularity of f implies a faster decay of the Fourier coefficients, and therefore a convergence in stronger norms of the Fourier series.

Proposition 5.8. *Assume that $f \in C^1([-\pi, \pi])$ and that $f(-\pi) = f(\pi)$. Then the Fourier series of f converges uniformly to f in $[-\pi, \pi]$.*

Proof. We first notice that $\widetilde{f}$ in (5.4) is Lipschitz continuous, so that by Proposition 5.3 we have

$$f(x) = \frac{1}{2} a_0 + \sum_{k=1}^{\infty} (a_k \cos kx + b_k \sin kx) \qquad \forall x \in [-\pi, \pi].$$

Let us consider the Fourier series of the derivative f' of f,

$$\sum_{k=1}^{\infty} (a_k' \cos kx + b_k' \sin kx) \qquad x \in [-\pi, \pi],$$

where, for $k \geq 1$ integer,

$$a_k' = \frac{1}{\pi} \int_{-\pi}^{\pi} f'(y) \cos ky \, dy, \qquad b_k' = \frac{1}{\pi} \int_{-\pi}^{\pi} f'(y) \sin ky \, dy. \tag{5.13}$$

Notice that $a_0' = 0$ because $f(-\pi) = f(\pi)$ implies that the mean value of f' on $(-\pi, \pi)$ is 0. As easily checked through an integration by parts (using again the fact that $f(-\pi) = f(\pi)$), we have $a_k' = k b_k$ and $b_k' = -k a_k$. Then, by the Bessel inequality it follows that

$$\sum_{k=1}^{\infty} k^2 (a_k^2 + b_k^2) = \sum_{k=1}^{\infty} (a_k')^2 + (b_k')^2 \leq \frac{1}{\pi} \int_{-\pi}^{\pi} |f'(x)|^2 \, dx < \infty. \quad (5.14)$$

Therefore the Fourier series of f is totally convergent in $C([-\pi, \pi])$ and therefore uniformly convergent. We have indeed

$$\sum_{k=1}^{\infty} \max_{x \in [-\pi, \pi]} |a_k \cos kx + b_k \sin kx|$$

$$\leq \sum_{k=1}^{\infty} (|a_k| + |b_k|)$$

$$\leq \left(\sum_{k=1}^{\infty} k^2 (|a_k| + |b_k|)^2 \right)^{1/2} \left(\sum_{k=1}^{\infty} k^{-2} \right)^{1/2} < \infty. \qquad \square$$

Exercises

5.1 Check that the trigonometric system (5.1) is orthogonal.

5.2 Let E be a Banach space. Show that any totally convergent series $\sum_n x_n$, with $(x_n) \subset E$, is convergent. Moreover,

$$\left\| \sum_{n=0}^{\infty} x_n \right\| \leq \sum_{n=0}^{\infty} \|x_n\|. \qquad (5.15)$$

Hint: estimate $\| \sum_0^N x_n - \sum_0^M x_n \|$ with the triangle inequality.

5.3 Prove that the following systems on $L^2(0, \pi)$ are orthonormal and complete

$$\sqrt{\frac{2}{\pi}} \sin kx, \quad k \geq 1,$$

and

$$\frac{1}{\sqrt{\pi}}; \sqrt{\frac{2}{\pi}} \cos kx, \quad k \geq 1.$$

5.4 Show that

$$e_k(x) := \frac{1}{\sqrt{2\pi}} e^{ikx}, \quad k \in \mathbb{Z}$$

is a complete orthonormal system in $L^2((-\pi, \pi); \mathbb{C})$. *Hint:* in order to show completeness, consider first the cases where f is real-valued or if is real-valued.

5.5 Let (e_k) be as in Exercise 5.4. Using the Parseval identity show that

$$\int_{-\pi}^{\pi} |f(x)|^2 \, dx = \frac{1}{2\pi} \sum_{k \in \mathbb{Z}} \left(\int_{-\pi}^{\pi} f(x) e^{-ikx} \, dx \right)^2 \qquad \forall f \in L^2 \left((-\pi, \pi); \mathbb{C} \right).$$

5.6 Let $f \in L^2 \left((-\pi, \pi); \mathbb{C} \right)$ and let $S_N f = \sum_{-N}^{N} \langle f, e_k \rangle e_k$, with $N \geq 1$, be the Fourier sums corresponding to the complete orthonormal system in Exercise 5.4. Show that

$$f(x) - S_N f(x) = \int_{-\pi}^{\pi} G_N(x - y)(f(x) - f(y)) \, dy$$

with

$$G_N(z) := \frac{\sin((N + 1/2)z)}{\sin(z/2)}.$$

Hint: use the identities $\sum_0^N e^{iky} = \sum_0^N (e^{iy})^k = (e^{i(N+1)z} - 1)/(e^{iy} - 1)$.

5.7 Arguing as in Remark 5.7, show that $\sum_1^\infty k^{-4} = \pi^4/90$. *Hint:* consider the function $f(x) = x^2$.

5.8 *Chebyschev* polynomials C_n in $L^2(a, b)$, with (a, b) bounded interval, are the ones obtained by applying the Gram-Schmidt procedure to the vectors $1, x$, $x^2, x^3, \ldots$. They are also called *Legendre* polynomials when $(a, b) = (-1, 1)$.

(a) Compute explicitly the first three Legendre polynomials.
(b) Show that $\{C_n\}_{n \in \mathbb{N}}$ is a complete orthonormal system. *Hint:* use the density of polynomials in $C([a, b])$.
(c) ★ Show that the n-th Legendre polynomial P_n is given by

$$P_n(x) = \sqrt{\frac{2n + 1}{2}} \, \frac{1}{2^n n!} \frac{d^n}{d^n x} (x^2 - 1)^n.$$

5.9 Let $f \in C^m \left([-\pi, \pi]; \mathbb{C} \right)$ with $f^{(j)}(-\pi) = f^{(j)}(\pi)$ for all $j = 0, \ldots, m - 1$. Show that $c_k^{(m)}$, the k-th Fourier coefficient of $f^{(m)}$ is linked to c_k, the k-th Fourier coefficient of f, by $c_k^{(m)} = (ik)^m c_k$.

Chapter 6
Operations on measures

In this chapter we collect many useful tools in Analysis and Probability that will be widely used in the following chapters. We will study the product of measures (both finite and countable), the product of measures by L^1 functions, the Radon–Nikodým theorem, the convergence of measures on the real line $\mathbb{R}$ and the Fourier transform.

6.1. The product measure and Fubini–Tonelli theorem

Let $(X, \mathscr{F})$ and $(Y, \mathscr{G})$ be measurable spaces. Let us consider the product space $X \times Y$. A set of the form $A \times B$, where $A \in \mathscr{F}$ and $B \in \mathscr{G}$, is called a *measurable rectangle*. We denote by $\mathscr{R}$ the family of all measurable rectangles. $\mathscr{R}$ is obviously a π–system. The σ–algebra generated by $\mathscr{R}$ is called the *product σ–algebra* of $\mathscr{F}$ and $\mathscr{G}$. It is denoted by $\mathscr{F} \times \mathscr{G}$.

Given σ–finite measures μ in $(X, \mathscr{F})$ and ν in $(Y, \mathscr{G})$, we are going to define the *product measure* $\mu \times \nu$ in $(X \times Y, \mathscr{F} \times \mathscr{G})$.

First, for any $E \in \mathscr{F} \times \mathscr{G}$ we define the *sections* of E, setting for $x \in X$ and $y \in Y$,

$$E_x := \{y \in Y \,:\, (x, y) \in E\}, \qquad E^y := \{x \in X \,:\, (x, y) \in E\}.$$

Proposition 6.1. *Assume that μ and ν are σ–finite and let $E \in \mathscr{F} \times \mathscr{G}$. Then the following statements hold.*

(i) $E_x \in \mathscr{G}$ for all $x \in X$ and $E^y \in \mathscr{F}$ for all $y \in Y$.
(ii) The functions

$$x \mapsto \nu(E_x), \qquad y \mapsto \mu(E^y),$$

are $\mathscr{F}$–measurable and $\mathscr{G}$–measurable respectively. Moreover,

$$\int_X \nu(E_x)\,d\mu(x) = \int_Y \mu(E^y)\,d\nu(y). \qquad (6.1)$$

Proof. Assume first that $E = A \times B$ is a measurable rectangle. Then, if $(x, y) \in X \times Y$ we have

$$E_x = \begin{cases} B & \text{if } x \in A \\ \varnothing & \text{if } x \notin A, \end{cases} \qquad E^y = \begin{cases} A & \text{if } y \in B \\ \varnothing & \text{if } y \notin B. \end{cases}$$

Consequently,

$$v(E_x) = \mathbb{1}_A(x)v(B), \qquad \mu(E^y) = \mathbb{1}_B(y)\mu(A),$$

so that (6.1) clearly holds.

Now, let $\mathscr{D}$ be the family of all $E \in \mathscr{F} \times \mathscr{G}$ such that (i) is fulfilled. Clearly, $\mathscr{D}$ is a Dynkin system including the π–system $\mathscr{R}$. Therefore, (i) follows from the Dynkin theorem.

Now, if both μ are v are finite, let $\mathscr{D}$ be the family of all $E \in \mathscr{F} \times \mathscr{G}$ such that (ii) is fulfilled. Clearly, $\mathscr{D}$ is a Dynkin system including the π–system $\mathscr{R}$ (stability under complement follows by the identities $v((E^c)_x) = v(Y) - v(E_x)$ and $\mu((E^c)^y) = \mu(X) - \mu(E^y)$). Therefore, (ii) follows from the Dynkin theorem as well.

In the general σ–finite case we argue by approximation: if $E \in \mathscr{F} \times \mathscr{G}$, $\mathscr{F} \ni X_h \uparrow X$ and $\mathscr{G} \ni Y_h \uparrow Y$ satisfy $\mu(X_h) < \infty$ and $v(Y_h) < \infty$, we define the finite measures

$$\mu_h(A) = \mu(A \cap X_h), \qquad v_h(B) = v(B \cap Y_h)$$

to obtain that $x \mapsto v_h(E_y)$ is $\mathscr{F}$–measurable and $y \mapsto \mu_h(E_x)$ is $\mathscr{G}$–measurable for all $E \in \mathscr{E} \times \mathscr{G}$. Passing to the limit as $h \to \infty$ in the identity

$$\int_{X_h} v_h(E_x)\,d\mu(x) = \int_X v_h(E_x)\,d\mu_h(x) = \int_Y \mu_h(E^y)\,dv_h(y)$$
$$= \int_{Y_h} \mu_h(E^y)\,dv(y)$$

the continuity properties of measures and integrals give (6.1) as well. $\square$

Theorem 6.2 (Product measure). *If μ and v are σ–finite, there exists a unique measure λ in $(X \times Y, \mathscr{F} \times \mathscr{G})$ satisfying*

$$\lambda(A \times B) = \mu(A)v(B) \qquad \text{for all } A \in \mathscr{F},\ B \in \mathscr{G}.$$

The measure λ is σ-finite and denoted by $\mu \times v$. Furthermore $\mu \times v$ is finite (resp. a probability measure) if both μ and v are finite (resp. probability measures).

Proof. Existence is easy: we set

$$\lambda(E) = \int_X \nu(E_x)\, d\mu(x) = \int_Y \mu(E^y)\, d\nu(y), \qquad E \in \mathscr{F} \times \mathscr{G}. \quad (6.2)$$

Using the continuity and additivity properties of the integral, it is immediate to check that λ is a measure on $(X \times Y, \mathscr{F} \times \mathscr{G})$. In the case of σ–finite measures, uniqueness follows by the the coincidence criterion for positive measures stated in Proposition 1.15: indeed, the value of the product measure is uniquely determined on the π–system $\mathscr{K}$ made by rectangles $A \times B$ with $\mu(A)$ and $\nu(B)$ finite, and thanks to the σ–finiteness assumption there exist $E_n = A_n \times B_n \in \mathscr{K}$ with $E_n \uparrow X \times Y$. $\qquad\square$

Corollary 6.3. *Let $E \in \mathscr{F} \times \mathscr{G}$ be such that $\mu \times \nu(E) = 0$. Then $\mu(E^y) = 0$ for ν–almost all $y \in Y$ and $\nu(E_x) = 0$ for μ–almost all $x \in X$.*

Proof. It follows directly from (6.2). $\qquad\square$

We consider here the measure space $(X \times Y, \mathscr{F} \times \mathscr{G}, \lambda)$, where $\lambda = \mu \times \nu$ and μ and ν are σ–finite.

Theorem 6.4 (Fubini–Tonelli). *Let $F \colon X \times Y \to [0, +\infty]$ be a $\mathscr{F} \times \mathscr{G}$–measurable map. Then the following statements hold.*

(i) *For any $x \in X$ (respectively $y \in Y$), the function $y \mapsto F(x, y)$ (respectively $x \mapsto F(x, y)$) is $\mathscr{G}$–measurable (resp. $\mathscr{F}$–measurable).*
(ii) *The functions*

$$x \mapsto \int_Y F(x, y)\, d\nu(y), \qquad y \mapsto \int_X F(x, y)\, d\mu(x)$$

are respectively $\mathscr{F}$–measurable and $\mathscr{G}$–measurable.
(iii) *We have*

$$\int_{X \times Y} F(x, y)\, d\lambda(x, y) = \int_X \left[\int_Y F(x, y)\, d\nu(y) \right] d\mu(x)$$
$$= \int_Y \left[\int_X F(x, y)\, d\mu(x) \right] d\nu(y). \quad (6.3)$$

Proof. Assume first that $F = \mathbb{1}_E$, with $E \in \mathscr{F} \times \mathscr{G}$. Then we have

$$F(x, y) = \mathbb{1}_{E_x}(y), \quad x \in X, \qquad F(x, y)(x) = \mathbb{1}_{E^y}(x), \quad y \in Y,$$

so (i), (ii) and (iii) follow from Proposition 6.1. Consequently, by linearity, (i)–(iii) hold when F is a simple function. If F is general, it is enough to approximate it by a monotonically increasing sequence of simple functions and then pass to the limit using the monotone convergence theorem. $\qquad\square$

Remark 6.5 (The definition of integral revisited). We noticed in Remark 2.13 that the integral of nonnegative functions can also be defined without using the archimedean integral, by considering minorant simple functions. If we follow this approach, the identity that we used to define the integral can be derived by applying the Fubini–Tonelli theorem to the subgraph

$$E := \{(x, t) \in X \times \mathbb{R} : 0 < t < f(x)\},$$

with the product measure $\mu \times \lambda$, λ being the Lebesgue measure. Indeed, it is not difficult to show that E is $\mathscr{F} \times \mathscr{B}(\mathbb{R})$–measurable whenever f is $\mathscr{F}$-measurable, so that

$$\int_0^\infty \mu(\{f > t\})\, dt = \int_0^\infty \mu(E^t)\, dt = \mu \times \lambda(E) = \int_X \lambda(E_x)\, d\mu(x)$$

$$= \int_X f(x)\, d\mu(x).$$

Of course, splitting F in positive and negative parts, also the case of extended real valued maps can be considered:

Corollary 6.6. *Let $F : X \times Y \to [-\infty, +\infty]$ be a $\mathscr{F} \times \mathscr{G}$–measurable map. Then F is $\mu \times \nu$–integrable if and only if:*

(i) *for μ–a.e. $x \in X$ the function $y \mapsto F(x, y)$ is ν–integrable;*
(ii) *the function $x \mapsto \int_Y |F(x, y)|\, d\nu(y)$ is μ–integrable.*

If these conditions hold, we have

$$\int_{X \times Y} F(x, y)\, d(\mu \times \nu)(x, y) = \int_X \left[\int_Y F(x, y)\, d\nu(y) \right] d\mu(x). \quad (6.4)$$

Notice that, strictly speaking, the function in (ii) is defined only out of a μ–negligible set; by μ–integrability of it we mean μ–integrability of any $\mathscr{F}$–measurable extension of it (for instance we may set it equal to 0 wherever $\int_Y |F(x, y)|\, d\nu(y)$ is not finite).

Remark 6.7 (Finite products). The previous constructions extend without any difficulty to finite products of measurable spaces $(X_i, \mathscr{F}_i)$. Namely, the product σ-algebra $\mathscr{F} := \bigtimes_i^n \mathscr{F}_i$ in the cartesian product $X := \bigtimes_1^n X_i$ is generated by the rectangles

$$\{A_1 \times \cdots \times A_n \ : \ A_i \in \mathscr{F}_i,\ 1 \leq i \leq n\}.$$

Furthermore, if μ_i are σ–finite measures in $(X_i, \mathscr{F}_i)$, integrals with respect to the product measure $\mu = \underset{1}{\overset{n}{\times}} \mu_i$ are defined by

$$\int_X F(x)\, d\mu(x) = \int_{X_1} \int_{X_2} \cdots \int_{X_n} F(x_1, \ldots, x_n)\, d\mu_n(x_n) \cdots d\mu_2(x_2)\, d\mu_1(x_1),$$

and any permutation in the order of the integrals would produce the same result. Finally, the product measure is uniquely determined, in the σ–finite case, by the product rule

$$\mu(A_1 \times \cdots \times A_n) = \prod_{i=1}^{n} \mu_i(A_i) \qquad A_i \in \mathscr{F}_i, \ 1 \le i \le n.$$

It is also not hard to show that the product is associative, both at the level of σ–algebras and measures, see Exercise 6.1.

6.2. The Lebesgue measure on $\mathbb{R}^n$

This section is devoted to the construction, the characterization and the main properties of the Lebesgue measure in $\mathbb{R}^n$, *i.e.* the length measure in $\mathbb{R}^1$, the area measure in $\mathbb{R}^2$, the volume measure in $\mathbb{R}^3$ and so on.

Definition 6.8 (Lebesgue measure in $\mathbb{R}^n$). Let us consider the measure space $(\mathbb{R}, \mathscr{B}(\mathbb{R}), \mathscr{L}^1)$, where $\mathscr{L}^1$ is the Lebesgue measure on $(\mathbb{R}, \mathscr{B}(\mathbb{R}))$. Then, we can define the measure space $(\mathbb{R}^n, \underset{i=1}{\overset{n}{\times}} B(\mathbb{R}), \mathscr{L}^n)$ with $\mathscr{L}^n :=$ $\underset{1}{\overset{n}{\times}} \mathscr{L}^1$. We say that $\mathscr{L}^n$ is the Lebesgue measure on $\mathbb{R}^n$.

Since (see Exercise 6.2)

$$\mathscr{B}(\mathbb{R}^n) = \underset{i=1}{\overset{n}{\times}} \mathscr{B}(\mathbb{R}),$$

we can equivalently consider $\mathscr{L}^n$ as a measure in $(\mathbb{R}^n, \mathscr{B}(\mathbb{R}^n))$, forgetting its construction as a product measure (indeed, there exist alternative and direct constructions of $\mathscr{L}^n$ independent of the concept of product measure).

As in the one-dimensional case, we will keep using the classical notation

$$\int_E f(x)\, dx := \int_{\mathbb{R}^n} f \mathbb{1}_E \, d\mathscr{L}^n \qquad E \in \mathscr{B}(\mathbb{R}^n),\ f : \mathbb{R}^n \to \mathbb{R} \text{ Borel}$$

for integrals with respect to Lebesgue measure $\mathscr{L}^n$ (or Riemann integrals in more than one independent variable).

In the computation of Lebesgue integrals, a particular role is sometimes played by the dimensional constant $\omega_n = \mathscr{L}^n(B(0, 1))$ (so that $\omega_1 = 2$, $\omega_2 = \pi$, $\omega_3 = 4\pi/3,\ldots$). A general formula for the computation of ω_n can be given using Euler's Γ function:

$$\Gamma(z) := \int_0^\infty t^{z-1} e^{-t}\, dt \qquad\qquad z > 0.$$

Indeed, we have

$$\omega_n = \frac{\pi^{n/2}}{\Gamma(\frac{n}{2} + 1)}. \tag{6.5}$$

A proof of this formula, based on the identity $\Gamma(z + 1) = z\Gamma(z)$ (which gives also $\Gamma(n) = (n-1)!$ for $n \geq 1$ integer) is proposed in Exercise 6.7.

We are going to show that $\mathscr{L}^n$ is invariant under translations and rotations. For this we need some notation. For any $a \in \mathbb{R}^n$ and any $\delta > 0$ we set

$$Q(a, \delta) := \left\{ x \in \mathbb{R}^n : a_i \leq x_i < a_i + \delta, \quad \forall\, i = 1, \ldots, n \right\}$$
$$= \underset{i=1}{\overset{n}{\times}} [a_i, a_i + \delta).$$

$Q(a, \delta)$ is called the δ–*box with corner at* a. For all $N \in \mathbb{N}$ we consider the family

$$\mathscr{Q}_N = \{Q(2^{-N}k, 2^{-N}) : k = (k_1, \ldots, k_n) \in \mathbb{Z}^n\}.$$

It is also clear that each box in $\mathscr{Q}_N$ is Borel and that its Lebesgue measure is 2^{-nN}. Now we set

$$\mathscr{Q} = \bigcup_{N=0}^\infty \mathscr{Q}_N.$$

It is clear that all boxes in $\mathscr{Q}_N$ are mutually disjoint and that their union is $\mathbb{R}^n$. Furthermore, if $N < M$, $Q \in \mathscr{Q}_N$ and $Q' \in \mathscr{Q}_M$, then either $Q' \subset Q$ or $Q \cap Q' = \varnothing$. If follows that if Q, $Q' \in \mathscr{Q}$ intersect, then one of the two sets is contained in the other one.

Lemma 6.9. *Let U be a non empty open set in $\mathbb{R}^n$. Then U is the disjoint union of boxes in $\mathscr{Q}$.*

Proof. For any $x \in U$, let $Q_x \in \mathscr{Q}$ be the biggest box such that $x \in Q_x \subset U$. This box is uniquely defined: indeed, fix an x; for any m there is only one box $Q_{x,m} \in \mathscr{Q}_m$ such that $x \in Q_{x,m}$; moreover, since U is open, for m large enough $Q_{x,m} \subset U$; we can then define $Q_x = Q_{x,\tilde{m}}$ where $\tilde{m}$ is the smallest integer m such that $Q_{x,m} \subset U$.

This family $\{Q_x\}_{x \in U}$ is a partition of U, that is, for any x, $y \in U$, either $Q_x = Q_y$ or $Q_x \cap Q_y = \varnothing$; indeed, if we suppose that $Q_x \cap Q_y \neq \varnothing$, then one of the two boxes is contained in the other, say $Q_x \subset Q_y$. This leads to $x \in Q_x \subset Q_y \subset U$, contradicting the definition of Q_x unless $Q_x = Q_y$. $\qquad\square$

From Lemma 6.9 it follows easily that the σ–algebra generated by $\mathscr{Q}$ coincides with $\mathscr{B}(\mathbb{R}^n)$.

Proposition 6.10 (Properties of the Lebesgue measure). *The following statements hold.*

(i) *(translation invariance) For any $E \in \mathscr{B}(\mathbb{R}^n)$, $x \in \mathbb{R}^n$ we have $\mathscr{L}^n(E + x) = \mathscr{L}^n(E)$, where*

$$E + x = \{y + x : \ y \in \mathbb{R}^n\}.$$

(ii) *If μ is a translation invariant measure on $(\mathbb{R}^n, \mathscr{B}(\mathbb{R}^n))$ such that $\mu(K) < \infty$ for any compact set K, there exists a number $C_\mu \geq 0$ such that*

$$\mu(E) = C_\mu \mathscr{L}^n(E) \qquad \forall\, E \in \mathscr{B}(\mathbb{R}^n).$$

(iii) *(rotation invariance) For any orthogonal matrix $R \in L(\mathbb{R}^n; \mathbb{R}^n)$ we have*
$$\mathscr{L}^n(R(E)) = \mathscr{L}^n(E) \qquad \forall\, E \in \mathscr{B}(\mathbb{R}^n).$$

(iv) *For any $T \in L(\mathbb{R}^n; \mathbb{R}^n)$ we have*

$$\mathscr{L}^n(T(E)) = |\det T| \mathscr{L}^n(E) \qquad \forall\, E \in \mathscr{B}(\mathbb{R}^n).$$

Proof. Fix $x \in \mathbb{R}^n$. The measures $\mathscr{L}^n(E)$ and $\mathscr{L}^n(E + x)$ coincide on the π–system of boxes; thanks to Lemma 6.9, this π–system generates the Borel σ–algebra, so that the coincidence criterion for measures stated in Proposition 1.15 gives that $\mathscr{L}^n(E) = \mathscr{L}^n(E + x)$ for all Borel sets E. Let us prove (ii). Let $Q_0 \in \mathscr{Q}_0$ and set $C_\mu = \mu(Q_0)$. Since Q_0 is included in a compact set, we have $C_\mu < \infty$. Since μ is translation invariant, all boxes in $\mathscr{Q}_0$ have the same μ measure. Now, let $Q_N \in \mathscr{Q}_N$. Since Q_0 is the disjoint union of 2^{-nN} boxes in $\mathscr{Q}_N$ which have all the same μ measure (again by the translation invariance) we have that

$$\mu(Q_N) = C_\mu \mathscr{L}^n(Q_N).$$

So, Lemma 6.9 gives that $\mu(A) = C_\mu \mathscr{L}^n(A)$ for any open set, and therefore for any Borel set.

Let us now prove (iii). By the translation invariance of $\mathscr{L}^n$, the measure $\mu(E) = \mathscr{L}^n(R(E))$ is easily seen to be translation invariant (because $R(E+z) = R(E) + R(z)$), hence $\mathscr{L}^n(R(E)) = C\mathscr{L}^n(E)$ for some contant C. We can identify the constant C choosing E equal to the unit ball, finding $C = 1$.

Finally, let us prove (iv). By polar decomposition we can write $T = R \circ S$ with $S = \sqrt{T^* \circ T}$ symmetric and nonnegative definite, and R orthogonal. Notice that on one hand $|\det T| = \det S$ (because $\det R \in \{-1, 1\}$) and on the other hand, by (iii) we have

$$\mathscr{L}^n(T(E)) = \mathscr{L}^n(R(S(E))) = \mathscr{L}^n(S(E)).$$

Hence, it suffices to show that $\mathscr{L}^n(S(E)) = \det S \, \mathscr{L}^n(E)$ for any symmetric and nonnegative definite matrix S. By the translation invariance of $\mathscr{L}^n(S(E))$ there exists a constant C such that $\mathscr{L}^n(S(E)) = C\mathscr{L}^n(E)$ for any Borel set E. In this case we can identify the constant C choosing as E a suitable n-dimensional cube: denoting by (e_i) an orthonormal basis of eigenvectors of S, with eigenvalues $\alpha_i \geq 0$ (whose product is $\det S$), choosing

$$E = \left\{ \sum_{i=1}^{n} c_i e_i \ : \ |c_i| \leq 1 \right\}, \ \text{ so that } S(E) = \left\{ \sum_{i=1}^{n} \alpha_i c_i e_i \ : \ |c_i| \leq 1 \right\},$$

the rotation invariance of $\mathscr{L}^n$ gives $\mathscr{L}^n(E) = 1$ and $\mathscr{L}^n(S(E)) = \alpha_1 \cdots \alpha_n$. Therefore $C = \det S$ and the proof is complete. $\qquad \square$

6.3. Countable products

We are here concerned with a sequence $(X_i, \mathscr{F}_i, \mu_i)$, $i = 1, 2, \ldots$, of probability spaces. We denote by X the product space

$$X := \underset{k=1}{\overset{\infty}{\times}} X_k$$

and by $x = (x_k)$ the generic element of X.

We are going to define a σ–algebra of subsets of X. Let us first introduce the cylindrical sets in X. A *cylindrical set* $I_{n,A}$ is a set of the following form

$$I_{n,A} = \{x \ : \ (x_1, \ldots, x_n) \in A\},$$

where $n \geq 1$ is an integer and $A \in \underset{1}{\overset{n}{\times}} \mathscr{F}_k$. This representation is not unique; however, since

$$I_{n,A} = A \times \underset{k=n+1}{\overset{\infty}{\times}} X_k$$

we have that $I_{n,A} = I_{m,B}$ with $n < m$ implies $B = A \times X_{n+1} \times \cdots \times X_m$.

We denote by $\mathscr{C}$ the family of all cylindrical sets of X. Notice also that

$$I^c_{n,A} = I_{n,A^c},$$

so that $\mathscr{C}$ is stable under complement. If $I_{n,A}$ and $I_{m,B}$ belong to $\mathscr{C}$ we can assume by the previous remarks that $m = n$, so that $I_{n,A} \cup I_{n,B} = I_{n,A\cup B}$ belongs to $\mathscr{C}$. Therefore $\mathscr{C}$ is an algebra.

The σ–algebra generated by $\mathscr{C}$ is called the *product σ–algebra* of the σ–algebras $\mathscr{F}_i$. It is denoted by

$$\underset{k=1}{\overset{\infty}{\times}} \mathscr{F}_k.$$

Now we define a function μ on $\mathscr{C}$, setting

$$\mu(I_{n,A}) = \left(\underset{k=1}{\overset{n}{\times}} \mu_k \right)(A), \quad I_{n,A} \in \mathscr{C}. \tag{6.6}$$

This definition is well posed, again thanks to the fact that $I_{n,A} = I_{m,B}$ with $n < m$ when $B = A \times X_{n+1} \times \cdots \times X_m$. It is easy to check that μ is additive: indeed, if $I_{n,A}$ and $I_{m,B}$ are disjoint, using the previous remark we can assume with no loss of generality that $n = m$, and therefore the equality $\mu(I_{n,A} \cup I_{n,B}) = \mu(I_{n,A}) + \mu(I_{n,B})$ follows by

$$\left(\underset{k=1}{\overset{n}{\times}} \mu_k \right)(A \cup B) = \left(\underset{k=1}{\overset{n}{\times}} \mu_k \right)(A) + \left(\underset{k=1}{\overset{n}{\times}} \mu_k \right)(B).$$

Theorem 6.11. *The set function μ defined in (6.6) is σ–additive on $\mathscr{C}$ and therefore, by the Carathéodory theorem, it has a unique extension to a probability measure on $(X, \overset{\infty}{\underset{1}{\times}} \mathscr{F}_k)$ that is denoted by*

$$\underset{k=1}{\overset{\infty}{\times}} \mu_k$$

Proof. To prove the σ–additivity of μ it is enough to show the continuity of μ at $\varnothing$, or equivalently the implication

$$(E_j) \subset \mathscr{C}, \ (E_j) \text{ nonincreasing}, \ \mu(E_j) \geq \varepsilon_0 > 0 \Longrightarrow \bigcap_{n=1}^{\infty} E_j \neq \varnothing. \tag{6.7}$$

In the following we are given a nonincreasing sequence (E_j) on $\mathscr{C}$ such that $\mu(E_j) \geq \varepsilon_0 > 0$. To prove (6.7), we need some more notation. We set

$$X^{(n)} = \underset{k=n+1}{\overset{\infty}{\times}} X_k$$

and we define $\mu^{(n)}$ on cylindrical sets of $X^{(n)}$ as in (6.6). Then, we consider the *sections* of E_j defined as

$$E_j(x_1) = \left\{ x^{(1)} \in X^{(1)} : (x_1, x^{(1)}) \in E_j \right\}, \quad x_1 \in X_1.$$

$E_j(x_1)$ is a cylindrical subset of $X^{(1)}$ and by the Fubini theorem we have

$$\mu(E_j) = \int_{X_1} \mu^{(1)}(E_j(x_1)) \, d\mu_1(x_1) \geq \varepsilon_0 > 0, \qquad j \geq 1. \qquad (6.8)$$

Set now

$$F_{j,1} = \left\{ x_1 \in X_1 : \mu^{(1)}(E_j(x_1)) \geq \frac{\varepsilon_0}{2} \right\}, \quad j \geq 1.$$

Then $F_{j,1}$ is not empty and by (6.8) we have

$$\mu(E_j) = \int_{F_{j,1}} \mu^{(1)}(E_j(x_1)) \, d\mu_1(x_1) + \int_{F_{j,1}^c} \mu^{(1)}(E_j(x_1)) \, d\mu_1(x_1)$$

$$\leq \mu_1(F_{j,1}) + \frac{\varepsilon_0}{2}.$$

Therefore $\mu_1(F_{j,1}) \geq \varepsilon_0/2$ for all $j \geq 1$.

Obviously $(F_{j,1})$ is a nonincreasing sequence of subsets of X_1. Since μ_1 is σ–additive, it is continuous at 0. Therefore, there exists $\alpha_1 \in \bigcap_1^\infty F_{j,1}$ and so

$$\mu^{(1)}(E_j(\alpha_1)) \geq \frac{\varepsilon_0}{2}, \quad j \geq 1. \qquad (6.9)$$

Consequently we have

$$E_j(\alpha_1) \neq \varnothing, \quad j \geq 1. \qquad (6.10)$$

Now we iterate the procedure: for any $x_2 \in X_2$ we consider the section

$$E_j(\alpha_1, x_2) = \left\{ x^{(2)} \in X^{(2)} : (\alpha_1, x_2, x^{(2)}) \in E_j \right\}, \quad j \geq 1.$$

By the Fubini theorem we have

$$\mu^{(1)}(E_j(\alpha_1)) = \int_{X_2} \mu^{(2)}(E_j(\alpha_1, x_2)) \, d\mu_2(x_2). \qquad (6.11)$$

We set

$$F_{j,2} = \left\{ x_2 \in X_2 : \mu^{(2)}(E_j(\alpha_1, x_2)) \geq \frac{\varepsilon_0}{4} \right\}, \quad j \geq 1.$$

Then by (6.9) and (6.10) we have

$$\frac{\varepsilon_0}{2} \le \mu^{(1)}(E_j(\alpha_1)) = \int_{X_2} \mu^{(2)}(E_j(\alpha_1, x_2))\, d\mu_2(x_2)$$

$$= \int_{F_{j,2}} \mu^{(2)}(E_j(\alpha_1, x_2))\, d\mu_2(x_2) + \int_{[F_{j,2}]^c} \mu^{(2)}(E_j(\alpha_1, x_2))\, d\mu_2(x_2)$$

$$\le \mu_2(F_{j,2}) + \frac{\varepsilon_0}{4}.$$

Therefore $\mu_2(F_{j,2}) \ge \varepsilon_0/4$. Since $(F_{j,2})$ is nonincreasing and μ_2 is $\sigma-$additive, there exists $\alpha_2 \in X_2$ such that

$$\mu^2(E_j(\alpha_1, \alpha_2)) \ge \frac{\varepsilon_0}{4}, \quad j \ge 1,$$

and consequently we have

$$E_j((\alpha_1, \alpha_2)) \ne \varnothing. \tag{6.12}$$

Arguing in a similar way we see that there exists a sequence $(\alpha_k) \subset X$ such that

$$E_j(\alpha_1, \ldots, \alpha_n) \ne \varnothing, \quad \text{for all } j, n \ge 1, \tag{6.13}$$

where

$$E_j(\alpha_1, \ldots, \alpha_n) = \left\{ x \in X^{(n)} : (\alpha_1, \ldots, \alpha_n, x^{(n)}) \in E_j \right\}, \quad j, n \ge 1.$$

Since E_j are cylindrical, this easily implies that $(\alpha_n) \in \bigcap_1^\infty E_j$. Therefore $\bigcap_1^\infty E_j$ is not empty, as required. $\qquad\square$

Exercises

6.1 Let $(X_1, \mathscr{F}_1)$, $(X_2, \mathscr{F}_2)$, $(X_3, \mathscr{F}_3)$ be measurable spaces. Show that

$$(\mathscr{F}_1 \times \mathscr{F}_2) \times \mathscr{F}_3 = \mathscr{F}_1 \times (\mathscr{F}_2 \times \mathscr{F}_3).$$

If we are given measures μ_i in $\mathscr{F}_i$, $i = 1, 2, 3$, show also that $(\mu_1 \times \mu_2) \times \mu_3 = \mu_1 \times (\mu_2 \times \mu_3)$.

6.2 Let us consider the measurable spaces $(\mathbb{R}, \mathscr{B}(\mathbb{R}))$, $(\mathbb{R}^n, \mathscr{B}(\mathbb{R}^n))$. Show that

$$\mathscr{B}(\mathbb{R}^n) = \underset{i=1}{\overset{n}{\times}} \mathscr{B}(\mathbb{R}).$$

Hint: to show the inclusion $\subset$, use Lemma 6.9.

6.3 Let $\mathscr{L}_n$ be the $\sigma-$algebra of Lebesgue measurable sets in $\mathbb{R}^n$. Show that

$$\mathscr{L}_1 \times \mathscr{L}_1 \subsetneq \mathscr{L}_2.$$

Hint: to show the strict inclusion, consider the set $E = F \times \{0\}$, where $F \subset \mathbb{R}$ is not Lebesgue measurable.

6.4 Show that the product σ–algebra is also generated by the family of products $\times_1^\infty A_i$ where $A_i \in \mathscr{F}_i$ and $A_i \neq X_i$ only for finitely many i.

6.5 Writing properly $\mathscr{L}^3$ as a product measure, compute $\mathscr{L}^3(T)$, where

$$T = \left\{ (x, y, z) : \; x^2 + y^2 < r^2 \text{ and } y^2 + z^2 < r^2 \right\} .$$

6.6 [Computation of ω_n] Find a recursive formula linking ω_n to ω_{n-2}, and use it to show that $\omega_{2k} = \pi^k/k!$ and $\omega_{2k+1} = 2^{k+1}\pi^k/(2k+1)!!$, where $(2k+1)!!$ is the product of all odd integers between 1 and $2k+1$. *Hint:* use the Fubini–Tonelli theorem.

6.7 Use Exercise 6.6 and the identities $\Gamma(1) = 1$, $\Gamma(1/2) = \sqrt{\pi}$ and $\Gamma(z+1) = z\Gamma(z)$ to show (6.5).

6.8 Let μ and ν be σ–finite measures on $(X, \mathscr{F})$ and $(Y, \mathscr{G})$ respectively and let $\lambda = \mu \times \nu$. Let $\mathscr{E} = (\mathscr{F} \times \mathscr{G})_\lambda$, as defined in Definition 1.12, and let ζ be the extension of λ to $\mathscr{E}$. Show this version of the Fubini–Tonelli Theorem 6.4: for any $\mathscr{E}$–measurable function $F : X \times Y \to [0, +\infty]$ the following statements hold:

(i) for μ–a.e. $x \in X$ the function $y \mapsto F(x, y)$ is ν–measurable;

(ii) the function $x \mapsto \int_Y F(x, y)\, d\nu(y)$, set to zero at all points x such that $y \mapsto F(x, y)$ is not ν–integrable, is μ–measurable;

(iii) $\int_X \int_Y F(x, y)\, d\mu(x)\, d\mu(y) = \int_{X \times Y} F(x, y)\, d\zeta(x, y)$.

6.9 Using the notation of the Fubini-Tonelli theorem, let $X = Y = [0, 1]$, $\mathscr{F} = \mathscr{G} = \mathscr{B}([0, 1])$, let μ be the Lebesgue measure and let ν be the counting measure. Let $D = \{(x, x) \; : \; x \in [0, 1]\}$ be the diagonal in $X \times Y$; check that $\int_X \nu(D_x)\, d\mu(x) \neq \int_Y \mu(D^y)\, d\nu(y)$.

6.10 ★ Let (f_h) be converging to f in $L^1(X \times Y, \mu \times \nu)$. Show the existence of a subsequence $h(k)$ such that $f_{h(k)}(x, \cdot)$ converge to $f(x, \cdot)$ in $L^1(Y, \nu)$ for μ–a.e. $x \in X$. Show by an example that, in general, this property is not true for the whole sequence.

6.4. Comparison of measures

In this section we study some relations between measures in a measurable space $(X, \mathscr{F})$.

The first (immediate) one is the order relation: viewing measures as set functions, we say that $\mu \leq \nu$ if $\mu(B) \leq \nu(B)$ for all $B \in \mathscr{F}$. It is not hard to see that the space of measures endowed with this order relation is a complete lattice (see Exercise 6.13): in particular

$$\mu \vee \nu(B) = \sup \{\mu(A_1) + \nu(A_2) \; : \; A_1, A_2 \in \mathscr{F}, (A_1, A_2) \text{ partition of } B\}$$

and

$$\mu \wedge \nu(B) = \inf\{\mu(A_1) + \nu(A_2) \ : \ A_1, A_2 \in \mathscr{F}, (A_1, A_2) \text{ partition of } B\}.$$

Another relation between measures is linked to the concept of product of a function by a measure.

Definition 6.12. Let μ be a measure in $(X, \mathscr{F})$ and let $f \in L^1(X, \mathscr{F}, \mu)$ be nonnegative. We set

$$f\mu(B) := \int_B f \, d\mu \qquad \forall B \in \mathscr{F}. \tag{6.14}$$

It is immediate to check, using the additivity and the continuity properties of the integral, that $f\mu$ is a finite measure. Furthermore, the following simple rule provides a way for the computation of integrals with respect to $f\mu$:

$$\int_X h \, d(f\mu) = \int_X hf \, d\mu, \tag{6.15}$$

whenever h is $\mathscr{F}$–measurable and nonnegative (or hf is μ–integrable, see Exercise 6.11). It suffices to check the identity (6.15) on characteristic functions $h = \mathbb{1}_B$ (and in this case it reduces to (6.14)), and then for simple functions. The monotone convergence theorem then gives the general result.

Notice also that, by definition, $f\mu(B) = 0$ whenever $\mu(B) = 0$. We formalize this relation between measures in the next definition.

Definition 6.13 (Absolute continuity). Let μ, ν be measures in $\mathscr{F}$. We say that ν is absolutely continuous with respect to μ, and write $\nu \ll \mu$, if all μ–negligible sets are ν–negligible, *i.e.*

$$\mu(A) = 0 \qquad \Longrightarrow \qquad \nu(A) = 0.$$

For finite measures, the absolute continuity property can also be given in a (seemingly) stronger way, see Exercise 6.14.

The following theorem shows that absolute continuity of ν with respect to μ is not only necessary, but also sufficient to ensure the representation $\nu = f\mu$.

Theorem 6.14 (Radon–Nikodým). *Let μ and ν be finite measures on $(X, \mathscr{F})$ such that $\nu \ll \mu$. Then there exists a unique nonnegative $\rho \in L^1(X, \mathscr{F}, \mu)$ such that*

$$\nu(E) = \int_E \rho(x) \, d\mu(x) \qquad \forall E \in \mathscr{F}. \tag{6.16}$$

We are going to show a more general result, whose statement needs two more definitions. We say that a measure μ is *concentrated* on a $\mathscr{F}$–measurable set A if $\mu(X \setminus A) = 0$. For instance, the Dirac measure δ_a is concentrated on $\{a\}$, and the Lebesgue measure in $\mathbb{R}$ is concentrated on the irrational numbers, and $f\mu$ is concentrated (whatever μ is) on $\{f \neq 0\}$.

Definition 6.15 (Singular measures). Let μ, ν be measures in $(X, \mathscr{F})$. We say that μ is singular with respect to ν, and write $\mu \perp \nu$, if there exist disjoint $\mathscr{F}$–measurable sets A, B such that μ is concentrated on A and ν is concentrated on B.

The relation of singularity, as stated, is clearly symmetric. However, it can also be stated in a (seemingly) asymmetric way, by saying that $\mu \perp \nu$ if μ is concentrated on a ν–negligible set A (just take $B = A^c$ to see the equivalence with the previous definition).

Example 6.16. Let $X = \mathbb{R}$, $\mathscr{F} = \mathscr{B}(\mathbb{R})$, μ the Lebesgue measure on $(X, \mathscr{F})$ and $\nu = \delta_{x_0}$ the Dirac measure at $x_0 \in \mathbb{R}$. Then μ is concentrated on $A := \mathbb{R} \setminus \{x_0\}$, whereas ν is concentrated on $B := \{x_0\}$. So, μ and ν are singular.

Theorem 6.17 (Lebesgue). *Let μ and ν be measures on $(X, \mathscr{F})$, with μ σ–finite and ν finite. Then the following assertions hold.*

(i) *There exist two unique finite measures ν_a and ν_s on $(X, \mathscr{F})$ such that*

$$\nu = \nu_a + \nu_s, \qquad \nu_a \ll \mu, \qquad \nu_s \perp \mu. \qquad (6.17)$$

(ii) *There exists a unique $\rho \in L^1(X, \mathscr{F}, \mu)$ such that $\nu_a = \rho\mu$.*

(6.17) is called the *Lebesgue decomposition* of ν with respect to μ. The function ρ in (ii) is called the *density* of ν with respect to μ and it is sometimes denoted by

$$\rho := \frac{d\nu}{d\mu} \ .$$

Radon–Nikodým theorem simply follows by the Lebesgue theorem noticing that, in the case when $\nu \ll \mu$ the uniqueness of the decomposition gives $\nu_a = \nu$ and $\nu_s = 0$, so that $\nu = \nu_a = \rho\mu$.

Proof of Theorem 6.17. . We assume first that also μ is finite. Set $\lambda = \mu + \nu$ and notice that, obviously, $\mu \ll \lambda$ and $\nu \ll \lambda$. Define a linear functional F on $L^2(X, \mathscr{F}, \lambda)$ setting

$$F(\varphi) := \int_X \varphi(x) \, d\nu(x), \qquad \varphi \in L^2(X, \mathscr{F}, \lambda).$$

The functional F is well defined and bounded (and consequently continuous) since, in view of the Hölder inequality, we have

$$|F(\varphi)| \leq \int_X |\varphi(x)| \, d\nu(x) \leq \int_X |\varphi(x)| \, d\lambda(x) \leq [\lambda(X)]^{1/2} \, \|\varphi\|_{L^2(X,\mathscr{F},\lambda)}.$$

Now, thanks to the Riesz theorem, there exists a unique function $f \in L^2(X, \mathscr{F}, \lambda)$ such that

$$\int_X \varphi(x) \, d\nu(x) = \int_X f(x)\varphi(x) \, d\lambda(x) \qquad \forall \varphi \in L^2(X, \mathscr{F}, \lambda). \quad (6.18)$$

Setting $\varphi = \mathbb{1}_E$, with $E \in \mathscr{F}$, yields

$$\nu(E) = \int_E f(x) \, d\lambda(x) \geq 0,$$

which implies, by the arbitrariness of E, $f(x) \geq 0$, λ–a.e. and, in particular, both μ–a.e. and ν–a.e. In the sequel we shall assume, possibly modifying f in a λ–negligible set, and preserving the validity of (6.18), that $f \geq 0$ everywhere. By (6.18) it follows

$$\int_X \varphi(x)(1 - f(x)) \, d\nu(x) = \int_X f(x)\varphi(x) \, d\mu(x) \qquad \forall \varphi \in L^2(X, \mathscr{F}, \lambda).$$
$$(6.19)$$

Setting $\varphi = \mathbb{1}_E$, with $E \in \mathscr{F}$, yields

$$\int_E (1 - f(x)) \, d\nu(x) = \int_E f(x) \, d\mu(x) \geq 0$$

because $f \geq 0$. Thus, being E arbitrary, we obtain that $f(x) \leq 1$ for ν–a.e. $x \in X$. Set now

$$A := \{x \in X \ : \ 0 \leq f(x) < 1\}, \qquad B := \{x \in X \ : \ f(x) \geq 1\},$$

so that (A, B) is a $\mathscr{F}$–measurable partition of X, and

$$\nu_a(E) := \nu(E \cap A), \quad \nu_s(E) := \nu(E \cap B) \qquad \forall E \in \mathscr{F},$$

so that $\nu_a = \mathbb{1}_A \nu$ is concentrated on A, $\nu_s = \mathbb{1}_B \nu$ is concentrated on B and $\nu = \nu_a + \nu_s$.

Then, setting in (6.19) $\varphi = \mathbb{1}_B$, we see that

$$\mu(B) \leq \int_B f \, d\mu = \int_B (1 - f) \, d\nu = 0$$

because $f = 1$ ν–a.e. on B. It follows that ν_s is singular with respect to μ.

We show now that the existence of ρ such that $\nu_a = \rho\mu$. Heuristically, this can be obtained choosing in (6.19) the function $\varphi = (1 - f)^{-1}\mathbb{1}_{E\cap A}$, but since this function need not to be in $L^2(X, \mathscr{F}, \lambda)$ we argue by approximation: set in (6.19)

$$\varphi(x) = (1 + f(x) + \cdots + f^n(x))\mathbb{1}_{E\cap A}(x)$$

where $n \geq 1$ and $E \in \mathscr{F}$. Then we obtain

$$\int_{E\cap A} (1 - f^{n+1}(x))\, d\nu(x) = \int_{E\cap A} [f(x) + f^2(x) + \cdots + f^{n+1}(x)]\, d\mu(x).$$

Set $\rho(x) = 0$ for $x \in B$ and

$$\rho(x) := \lim_{n\to\infty} [f(x) + f^2(x) + \cdots + f^{n+1}(x)] = \frac{f(x)}{1 - f(x)}, \quad x \in A.$$

Then, by the monotone convergence theorem it follows that

$$\nu_a(E) = \nu(E \cap A) = \int_{E\cap A} \rho(x)\, d\mu(x) = \int_E \rho(x)\, d\mu(x).$$

Setting $E = X$ we see that $\rho \in L^1(X, \mathscr{F}, \mu)$, and the arbitrariness of E gives that $\nu_a = \rho\mu$.

Now we consider the case when μ is σ–finite. In this case there exists a sequence of pairwise disjoint sets $(X_n) \subset \mathscr{F}$ such that

$$X = \bigcup_{n=0}^{\infty} X_n \qquad \text{with } \mu(X_n) < \infty.$$

Let us apply Theorem 6.17 to the finite measures $\mu_n = \mathbb{1}_{X_n}\mu$, $\nu_n = \mathbb{1}_{X_n}\nu$. For any $n \in \mathbb{N}$ let $\nu_n = (\nu_n)_a + (\nu_n)_s = \rho_n\mu_n + (\nu_n)_s$ be the Lebesgue decomposition of ν_n with respect to μ_n. Now, set

$$\nu_a := \sum_{n=0}^{\infty} (\nu_n)_a, \qquad \nu_s := \sum_{n=0}^{\infty} (\nu_n)_s, \qquad \rho := \sum_{n=0}^{\infty} \rho_n\mathbb{1}_{X_n}.$$

Since

$$\sum_{n=0}^{k} (\nu_n)_a + (\nu_n)_s = \sum_{n=0}^{k} \nu_n = \mathbb{1}_{\cup_0^k X_n}\nu,$$

we can pass to the limit as $k \to \infty$ to obtain that ν_a and ν_s are finite measures, and $\nu = \nu_a + \nu_s$. Moreover, for any $E \in \mathscr{F}$ we have, using the monotone convergence theorem,

$$\nu_a(E) = \sum_{n=0}^{\infty} (\nu_n)_a(E) = \sum_{n=0}^{\infty} \int_E \rho_n(x)\, d\mu_n(x)$$

$$= \int_E \sum_{n=0}^{\infty} \rho_n(x) \mathbb{1}_{X_n}\, d\mu(x) = \int_E \rho(x)\, d\mu(x).$$

So, $\nu_a \ll \mu$, and setting $E = X$ we see that ρ is integrable with respect to μ. Finally, it is easy to see that $\nu_s \perp \mu$, because if we denote by $B_n \in \mathscr{F}$ μ–negligible sets where $(\nu_n)_s$ are concentrated, we have that ν_s is concentrated on the μ–negligible set $\cup_n B_n$.

Finally, let us prove the uniqueness of ν_a and ν_s: assume that

$$\nu = \nu_a + \nu_s = \nu_a' + \nu_s'$$

and let B, B' be μ–negligible sets where ν_s and ν_s' are respectively concentrated. Then, as $B \cup B'$ is μ–negligible and both ν_s and ν_s' are concentrated on $B \cup B'$, for any set $E \in \mathscr{F}$ we have

$$\nu_s(E) = \nu_s(E \cap (B \cup B')) = \nu(E \cap (B \cup B')) = \nu_s'(E \cap (B \cup B')) = \nu_s'(E).$$

It follows that $\nu_s = \nu_s'$ and therefore $\nu_a = \nu_a'$. $\qquad\square$

The interested reader can have a look at a different proof of Theorem 6.17 independent of Hilbert space theory, and based on three auxiliary variational principles; it turns out that the density f of ν^a is the maximizer in the problem

$$\sup \left\{ \int_X f\, d\mu : f\mu \leq \nu \right\}. \tag{6.20}$$

See Exercise 6.17 and Exercise 6.18 for more details.

Remark 6.18. If μ is not σ–finite then the Lebesgue decomposition does not hold in general. Consider for instance the case when $X = [0, 1]$, $\mathscr{F} = \mathscr{B}([0, 1])$, μ is the counting measure and $\nu = \mathscr{L}^1$. Then $\nu \ll \mu$ (as the only μ–negligible set is the empty set) but there is no $\rho : [0, 1] \to [0, \infty]$ satisfying

$$\nu(E) = \int_E \rho\, d\mu.$$

Indeed, this function should be μ-integrable and therefore it can be nonzero only in a set at most countable.

Exercises

6.11 Show that a $\mathscr{F}$–measurable function h is $f\mu$–integrable if and only if fh is μ–integrable.

6.12 Show that $(f\mu)\vee(g\mu) = (f\vee g)\mu$ and $(f\mu)\wedge(g\mu) = (f\wedge g)\mu$ whenever $f,\ g \in L^1(X, \mathscr{F}, \mu)$ are nonnegative.

6.13 Let $\{\mu_i\}_{i\in I}$ be a family of measures in $(X, \mathscr{F})$. Show that

$$\underline{\mu}(B) := \inf\left\{ \sum_{k=0}^{\infty} \mu_{i(k)}(B_k) \ :\ i : \mathbb{N} \to I, \right.$$

$$\left. (B_k) \text{ countable } \mathscr{F}\text{–measurable partition of } B \right\}$$

is the greatest lower bound of the family $\{\mu_i\}_{i\in I}$, *i.e.* $\underline{\mu} \le \mu_i$ for all $i \in I$ and it is the largest measure with this property. Show also that

$$\overline{\mu}(B) := \sup\left\{ \sum_{k=0}^{\infty} \mu_{i(k)}(B_k) \ :\ i : \mathbb{N} \to I, \right.$$

$$\left. (B_k) \text{ countable } \mathscr{F}\text{–measurable partition of } B \right\}$$

is the smallest upper bound of the family $\{\mu_i\}_{i\in I}$, *i.e.* $\overline{\mu} \ge \mu_i$ for all $i \in I$ and it is the smallest measure with this property.

6.14 Let $\mu,\ \nu$ be measures in $(X, \mathscr{F})$ with ν finite. Then $\nu \ll \mu$ if and only if for all $\varepsilon > 0$ there exists $\delta > 0$ such that

$$A \in \mathscr{F}, \ \mu(A) < \delta \qquad \Longrightarrow \qquad \nu(A) < \varepsilon.$$

6.15 Assume that $\nu \ll \mu$ and that $\nu \perp \mu$. Show that $\nu = 0$.

6.16 Assume that $\sigma \le \mu + \nu$ and that $\sigma \perp \nu$. Show that $\sigma \le \mu$.

6.17★ Prove Theorem 6.14 in the following two steps:

(1) Show that a maximizer f in (6.20) exists.
(2) Setting $\sigma = \nu - f\mu \ge 0$, σ satisfies

$$t > 0, \ \ B \in \mathscr{F}, \ \ t\mathbb{1}_B\mu \le \sigma \qquad \Longrightarrow \qquad \mu(B) = 0. \qquad (6.21)$$

Then, apply Exercise 6.18 to conclude that $\sigma \perp \mu$.

6.18★ ★ Let $\mu,\ \sigma$ be nonnegative finite measures satisfying (6.21). Show that $\sigma \perp \mu$. *Hint:* first show that

$$\inf\{\mu(A) \ :\ A \in \mathscr{F}, \ \ \sigma \text{ is concentrated on } A\}$$

has a solution A. Assuming by contradiction that $\mu(A) > 0$ (otherwise we are done), show that

$$\mathscr{F} \ni B \subseteq A, \ \ \mu(B) > 0 \qquad \Longrightarrow \qquad \sigma(B) > 0. \qquad (6.22)$$

Then, show that the numbers

$$\xi_h := \sup \left\{ \mu(B) : \ \mathscr{F} \ni B \subseteq A, \ \mathbb{1}_B \mu \geq 2^h \mathbb{1}_B \sigma \right\}$$

are infinitesimal as $h \to \infty$, that the supremum is attained at B_h, and that

$$\mu(C) \leq 2^h \sigma(C) \qquad \text{for all sets } C \subset A \setminus B_h. \tag{6.23}$$

Finally choose $t = 2^{-h}$, with h sufficiently large so that $\xi_h < \mu(A)$ and $B = A \setminus B_h$, to get a contradiction with (6.21).

6.5. Signed measures

Let $(X, \mathscr{F})$ be a measurable space. In this section we see how the concept of measure, still viewed as a set function, can be extended dropping the nonnegativity assumption on $A \mapsto \mu(A)$.

We recall that sequence $(E_i) \subset \mathscr{F}$ of pairwise disjoint sets such that $\bigcup_0^\infty E_i = E$ is called a countable $\mathscr{F}$*–measurable partition* of E.

Definition 6.19 (Signed measures and total variation). A *signed measure* μ in $(X, \mathscr{F})$ is a map $\mu : \mathscr{F} \to \mathbb{R}$ such that

$$\mu(E) = \sum_{i=0}^{\infty} \mu(E_i)$$

for all countable $\mathscr{F}$*–measurable partitions* (E_i) of E.

Notice that the series above is absolutely convergent by the arbitrariness of (E_i): indeed, if $\sigma : \mathbb{N} \to \mathbb{N}$ is a permutation, then $(E_{\sigma(i)})$ is still a partition of E, hence

$$\sum_{i=0}^{\infty} \mu(E_i) = \sum_{i=0}^{\infty} \mu(E_{\sigma(i)}).$$

This implies that the series is absolutely convergent.
Let μ be a signed measure. Then we define the *total variation* $|\mu|$ of μ as follows:

$$|\mu|(E) = \sup \left\{ \sum_{i=0}^{\infty} |\mu(E_i)| : \ (E_i) \ \mathscr{F}\text{–measurable partition of } E \right\},$$

$$E \in \mathscr{F}.$$

Proposition 6.20. *Let μ be a signed measure and let $|\mu|$ be its total variation. Then $|\mu|$ is a finite measure on $(X, \mathscr{F})$.*

Proof. It is immediate to check that $|\mu|$ is a nondecreasing set function.
Step 1. If A, $B \in \mathcal{F}$ are disjoint, we have

$$|\mu|(A \cup B) = |\mu|(A) + |\mu|(B).$$

Indeed, let $E = A \cup B$ and let (E_i) be a countable $\mathcal{F}$–measurable partition of E. Set

$$A_j = A \cap E_j, \quad B_j = B \cap E_j, \quad j \in \mathbb{N}.$$

Then (A_j) is a countable $\mathcal{F}$–measurable partition of A and (B_j) a countable $\mathcal{F}$–measurable partition of B and we have $E_j = A_j \cup B_j$. Moreover,

$$\sum_{j=0}^{\infty} |\mu(E_j)| \leq \sum_{j=1}^{\infty} |\mu(A_j)| + \sum_{j=0}^{\infty} |\mu(B_j)| \leq |\mu|(A) + |\mu|(B),$$

which yields $|\mu|(A \cup B) \leq |\mu|(A) + |\mu|(B)$.

Let us prove the converse inequality, assuming with no loss of generality that $|\mu|(A \cup B) < \infty$. Since both $|\mu|(A)$ and $|\mu|(B)$ are finite, for any $\varepsilon > 0$ there exist countable $\mathcal{F}$–measurable partitions (A_k^{ε}) of A and (B_k^{ε}) of B such that

$$\sum_{k=0}^{\infty} |\mu(A_k^{\varepsilon})| \geq |\mu|(A) - \frac{\varepsilon}{2}, \qquad \sum_{k=0}^{\infty} |\mu(B_k^{\varepsilon})| \geq |\mu|(B) - \frac{\varepsilon}{2}.$$

Since $(A_k^{\varepsilon}, B_k^{\varepsilon})$ is a countable $\mathcal{F}$–measurable partition of $A \cup B$, we have that

$$|\mu|(A \cup B) \geq \sum_{k=1}^{\infty} (|\mu(A_k^{\varepsilon})| + |\mu(B_k^{\varepsilon})|) \geq |\mu|(A) + |\mu|(B) - \varepsilon.$$

By the arbitrariness of ε we have $|\mu|(A \cup B) \geq |\mu|(A) + |\mu|(B)$.
Step 2. $|\mu|$ is σ–additive. Since $|\mu|$ is additive by Step 1, it is enough to show that $|\mu|$ is σ–subadditive, *i.e.* $|\mu(A)| \leq \sum_0^{\infty} |\mu|(A_i)$ whenever $(A_i) \subset \mathcal{F}$ is a partition of A. This can be proved arguing as in the first part of Step 1, *i.e.* building from a partition (E_j) of A partitions $(E_j \cap A_i)$ of all sets A_i.
Step 3. $|\mu|(X) < \infty$. Assume by contradiction that $|\mu|(X) = \infty$. Then we claim that

$$\text{there exists a partition } X = A \cup B \text{ such that}$$
$$|\mu(A)| \geq 1 \text{ and } |\mu|(B) = \infty. \tag{6.24}$$

By the claim the conclusion follows since we can use it to construct by recurrence (replacing X with B and so on), a disjoint sequence $(A_n) \subset \mathcal{F}$ such that $|\mu(A_n)| \geq 1$. Assume, to fix the ideas, that $\mu(A_n) \geq 1$ for infinitely many n, and denote by E the union of these sets: then, the σ–additivity of μ forces $\mu(E) = +\infty$, a contradiction. Analogously, if $\mu(A_n) \leq -1$ for infinitely many n, we find a set E such that $\mu(E) = -\infty$.

Let us prove (6.24). By the assumption $|\mu|(X) = \infty$ it follows the existence of a partition (X_n) of X such that

$$\sum_{n=0}^{\infty} |\mu(X_n)| > 2(1 + |\mu(X)|).$$

Then either the sum of those $\mu(X_n)$ which are nonnegative or the absolute value of the sum of those $\mu(X_n)$ which are nonpositive is greater than $1 + |\mu(X)|$. To fix the ideas, assume that for a subsequence $(X_{n(k)})$ we have $\mu(X_{n(k)}) \geq 0$ and

$$\sum_{k=0}^{\infty} \mu(X_{n(k)}) > 1 + |\mu(X)|.$$

Set $A = \bigcup_0^{\infty} X_{n(k)}$ and $B = A^c$. Then we have $|\mu(A)| > 1 + |\mu(X)|$ and

$$|\mu(B)| = |\mu(X) - \mu(A)| \geq |\mu(A)| - |\mu(X)| > 1.$$

Since

$$|\mu|(X) = |\mu|(A) + |\mu|(B) = \infty,$$

either $|\mu|(B) = +\infty$ or $|\mu|(A) = +\infty$. In the first case we are done, in the second one we exchange A and B. So, the claim is proved and the proof is complete. $\qquad\square$

Let μ be a signed measure on $(X, \mathcal{F})$. We define

$$\mu^+ := \frac{1}{2}\,(|\mu| + \mu), \qquad \mu^- := \frac{1}{2}\,(|\mu| - \mu),$$

so that

$$\mu = \mu^+ - \mu^- \qquad \text{and} \qquad |\mu| = \mu^+ + \mu^-. \tag{6.25}$$

The measure μ^+ (respectively μ^-) is called the *positive part* (respectively *negative part*) of μ and the first equation in (6.25) is called the *Jordan representation* of μ.

Remark 6.21. It is easy to check that Theorems 6.17 and 6.14 hold when v is a signed measure: it suffices to split it into its positive and negative part, see also Exercise 6.19.

The following theorem proves also that μ^+ and μ^- are singular, and provides a canonical representation of $\mu^\pm$ as suitable restrictions of $\pm\mu$.

Theorem 6.22 (Hahn decomposition). *Let μ be a signed measure on $(X, \mathscr{F})$ and let μ^+ and μ^- be its positive and negative parts. Then there exists a $\mathscr{F}$–measurable partition (A, B) of X such that*

$$\mu^+(E) = \mu(A\cap E) \quad and \quad \mu^-(E) = -\mu(B\cap E) \qquad \forall E \in \mathscr{F}. \quad (6.26)$$

Proof. Let us first notice that $\mu \ll |\mu|$. Thus, by the Radon–Nikodým theorem, there exists $h \in L^1(X, \mathscr{F}, |\mu|)$ such that

$$\mu(E) = \int_E h \, d|\mu| \qquad \forall E \in \mathscr{F}. \qquad (6.27)$$

Let us prove that $|h(x)| = 1$ for $|\mu|$–a.e. $x \in X$. Indeed, set

$$E_1 := \{x \in X : h(x) > 1\}, \qquad F_1 := \{x \in X : h(x) < -1\}$$

We first show that $|\mu|(E_1) = |\mu|(F_1) = 0$. Since we have

$$|\mu|(E_1) \geq \mu(E_1) = \int_{E_1} h \, d|\mu| \geq |\mu|(E_1),$$

and the second inequality is strict if $|\mu|(E_1) > 0$, we have that $|\mu|(E_1) = 0$. In a similar way one can prove that $|\mu|(F_1) = 0$, so that $|h| \leq 1$ $|\mu|$–a.e. in X. Now, let $r \in (0, 1)$ and set

$$G_r := \{x \in X : |h(x)| < r\}.$$

Let $(G_{r,k})$ be a countable $\mathscr{F}$–measurable partition of G_r. Then we have

$$|\mu(G_{r,k})| = \left| \int_{G_{r,k}} h \, d|\mu| \right| \leq \int_{G_{r,k}} |h| \, d|\mu| \leq r|\mu|(G_{r,k}).$$

Therefore

$$\sum_{k=0}^{\infty} |\mu(G_{r,k})| \leq r|\mu|(G_r),$$

which yields, by the arbitrariness of the partition of G_r, $|\mu|(G_r) \leq r|\mu|(G_r)$. Thus $|\mu|(G_r) = 0$ and letting $r \uparrow 1$ we obtain that $|\mu|(\{|h| <$

$1\}) = 0$. Hence, possibly modifying h in $|\mu|$–negligible set, we can assume with no loss of generality that h takes its values in $\{-1, 1\}$.

Now, to conclude the proof, we set

$$A := \{x \in X : h(x) = 1\}, \qquad B := \{x \in X : h(x) = -1\}.$$

Then for any $E \in \mathscr{F}$ we have

$$\mu^+(E) = \frac{1}{2}\,(|\mu|(E) + \mu(E)) = \frac{1}{2}\int_E (1 + h)d|\mu|$$

$$= \int_{E \cap A} h\,d|\mu| = \mu(E \cap A),$$

and

$$\mu^-(E) = \frac{1}{2}\,(|\mu|(E) - \mu(E)) = \frac{1}{2}\int_E (1 - h)d|\mu|$$

$$= -\int_{E \cap B} h\,d|\mu| = -\mu(E \cap B). \qquad \square$$

Exercises

6.19 Using the decomposition of ν in positive and negative part, show that Lebesgue decomposition is still possible when μ is σ–finite and ν is a signed measure. Using the Hahn decomposition extend this result to the case when even μ is a signed measure. Are these decompositions unique?

6.20 Show that $|f\mu| = |f|\mu$ for any $f \in L^1(X, \mathscr{E}, \mu)$.

6.6. Measures in $\mathbb{R}$

In this section we estabilish a 1-1 correspondence between finite Borel measures in $\mathbb{R}$ and a suitable class of nondecreasing functions. In one direction this correspondence is elementary, and based on the concept of *repartition function*.

Given a finite measure μ in $(\mathbb{R}, \mathscr{B}(\mathbb{R}))$, we call repartition function of μ the function $F : \mathbb{R} \to [0, +\infty)$ defined by

$$F(x) := \mu((-\infty, x]) \qquad x \in \mathbb{R}.$$

Notice that obviously[1] F is nondecreasing, right continuous, and satisfies

$$\lim_{x \to -\infty} F(x) = 0, \qquad \lim_{x \to +\infty} F(x) \in [0, +\infty). \qquad (6.28)$$

Moreover, F is continuous at x if and only if x is not an atom of μ.

---

[1] The arguments are similar to those used in Section 2.4.2, in connection with the properties of the function $t \mapsto \mu(\{\varphi > t\})$

The following result shows that this list of properties characterizes the functions that are repartition functions of some finite measure μ; in addition the measure is uniquely determined by its repartition function.

Theorem 6.23. *Let* $F : \mathbb{R} \to [0, +\infty)$ *be a nondecreasing and right continuous function satisfying* (6.28). *Then there exists a unique finite measure* μ *in* $(\mathbb{R}, \mathscr{B}(\mathbb{R}))$ *such that* F *is the repartition function of* μ.

Proof. The proof follows the same lines of the construction of the Lebesgue measure in Section 1.6, with a simplification due to the fact that we can also consider unbounded intervals (because we are dealing with finite measures). We set

$$\mathscr{I} := \{(a, b] \ : \ a \in [-\infty, +\infty), \ b \in \mathbb{R}, \ a < b\}$$

and denote by $\mathscr{A}$ the ring generated by $\mathscr{I}$: it consists, as it can be easily checked, of all finite disjoint unions of intervals in $\mathscr{I}$. We define, with the convention $F(-\infty) = 0$,

$$\mu((a, b]) := F(b) - F(a) \qquad \forall (a, b] \in \mathscr{I}. \tag{6.29}$$

This definition is justified by the fact that, if μ were a measure and F were its repartition function, (6.29) would be valid, because $(a, b] = (-\infty, b] \setminus (-\infty, a]$. Then we extend μ to $\mathscr{A}$ with the same mechanism used in the proof of Theorem 1.21, and check that μ is additive on $\mathscr{A}$. Also, the same argument used in that proof shows that μ is even $\sigma-$additive: in order to prove that $\mu(F) = \sum_i \mu(F_i)$ whenever F and all F_i belong to $\mathscr{A}$ one first reduces to the case when $F = (a, b]$ belongs to $\mathscr{I}$; then, one enlarges F_i to $F_i' \in \mathscr{A}$ with $\mu(F_i') < \mu(F_i) + \delta 2^{-i}$ and, using the fact that all intervals $[a', b]$ with $a' > a$ are contained in a finite union of the sets F_i', obtains

$$\mu((a', b]) \leq \sum_{i=0}^{\infty} \mu(F_i') \leq 2\delta + \sum_{i=0}^{\infty} \mu(F_i).$$

Letting first $\delta \downarrow 0$ and then $a' \downarrow a$ we obtain the $\sigma-$subadditivity property $\mu(F) \leq \sum_i \mu(F_i)$, and the opposite inequality follows by monotonicity.

By the Carathéodory theorem μ has a unique extension, that we still denote by μ, to $\mathscr{B}(\mathbb{R}) = \sigma(\mathscr{A})$. Setting $a = -\infty$ and letting b tend to $+\infty$ in the identity (6.29) we obtain that $\mu(\mathbb{R}) = F(+\infty) \in \mathbb{R}$. From (6.29) with $a = -\infty$ we obtain that the repartition function of μ is F. $\quad\square$

Given a nondecreasing and right continuous function F satisfying (6.28), the *Stieltjes* integral

$$\int_{\mathbb{R}} f \, dF$$

is defined as $\int f \, d\mu_F$, where μ_F is the finite measure built in the previous theorem. The notation dF is justified by the fact that, when $f = \sum_i z_i \mathbb{1}_{(a_i, b_i]}$, we have (by the very definition of μ_F)

$$\int_{\mathbb{R}} f \, dF = \int_{\mathbb{R}} f \, d\mu_F = \sum_i z_i (F(b_i) - F(a_i)).$$

This approximation of the Stieltjes integral will play a role in the proof of Theorem 6.28.

6.7. Convergence of measures on $\mathbb{R}$

In this section we study a notion of convergence for measures on the real line that is quite useful, both from the analytic and the probabilistic viewpoints.

Definition 6.24 (Weak convergence). Let (μ_h) be a sequence of finite measures on $\mathbb{R}$. We say that (μ_h) weakly converges to a finite measure μ on $\mathbb{R}$ if the repartition functions F_h of μ_h are pointwise converging to the repartition function F of μ on a co-countable set, *i.e.* if

$$\lim_{h \to \infty} \mu_h \left((-\infty, x]\right) = \mu \left((-\infty, x]\right) \text{ with at most countably many exceptions.}$$

$$(6.30)$$

Since the repartition function is right continuous, it is uniquely determined by (6.30). Then, since the measure is uniquely determined by its repartition function, we obtain that the weak limit, if exists, is unique. The following fundamental example shows why we admit at most countably many exceptions in the convergence of the repartition functions.

Example 6.25. [**Convergence to the Dirac mass**] Let $\rho \in C^{\infty}(\mathbb{R})$ be a nonnegative function such that $\int_{\mathbb{R}} \rho \, dx = 1$ (an important example is the Gauss function $(2\pi)^{-1/2} e^{-x^2/2}$). We consider the rescaled functions $\rho_h(x) = h\rho(hx)$ and the induced measures $\mu_h = \rho_h \mathcal{L}^1$, all probability measures. Then, it is immediate to check that μ_h weakly converge to δ_0: for $x > 0$ we have indeed

$$\mu_h \left((-\infty, x]\right) = \int_{-\infty}^{x} \rho_h(y) \, dy = \int_{-\infty}^{hx} \rho(y) \, dy \to 1$$

because $hx \to +\infty$ as $h \to +\infty$. An analogous argument shows that $\mu_h \left((-\infty, x]\right) \to 0$ for any $x < 0$. If ρ is even, at $x = 0$ we don't have pointwise convergence of the repartition functions: all the repartition functions F_h satisfy $F_h(0) = 1/2$, while $F(0) = 1$.

Weak convergence is a quite flexible tool, because it allows also an opposite behaviour, the approximation of *continuous* measures (*i.e.* with no atom) by *purely atomic* ones, see for instance Exercise 6.21.

From now on we will consider only, for the sake of simplicity, the case of weak convergence of *probability* measures. Before stating a compactness theorem for the weak convergence of probability measures, we introduce the following terminology.

Definition 6.26 (Tightness). We say that a family of probability measures $\{\mu_i\}_{i \in I}$ in $\mathbb{R}$ is *tight* if for any $\varepsilon > 0$ there exists a closed interval $J \subset \mathbb{R}$ such that

$$\mu_i(\mathbb{R} \setminus J) \le \varepsilon \qquad\qquad \forall i \in I.$$

Clearly any finite family of probability measures is tight. One can also check (see Exercise 6.24) that $\{\mu_i\}_{i \in I}$ is tight if and only if

$$\lim_{x \to -\infty} F_i(x) = 0, \quad \lim_{x \to +\infty} F_i(x) = 1 \qquad \text{uniformly with respect to } i \in I,$$

$$(6.31)$$

where F_i are the repartition functions of μ_i. Furthermore, (see Exercise 6.25) any weakly converging sequence is tight. Conversely, we have the following compactness result for tight sequences:

Theorem 6.27 (Compactness). *Let (μ_h) be a tight sequence of probability measures on $\mathbb{R}$. Then there exists a subsequence $(\mu_{h(k)})$ weakly converging to a probability measure μ.*

Proof. We denote by F_h the repartition functions of μ_h. By a diagonal argument we can find a subsequence $(F_{h(k)})$ pointwise converging on $\mathbb{Q}$. We denote by G the pointwise limit, obviously a nondecreasing function. We extend G by monotonicity setting

$$G(x) := \sup\{G(q) : q \in \mathbb{Q}, \ q \le x\} \qquad\qquad x \in \mathbb{R}$$

and let E be the co-countable set of the discontinuity points of G.

Let us check that $F_{h(k)}$ is pointwise converging to G on $\mathbb{R} \setminus E$: for $x \notin E$ we have indeed

$$\limsup_{k \to \infty} F_{h(k)}(x) \le \inf_{q \in \mathbb{Q}, \, q > x} \limsup_{k \to \infty} F_{h(k)}(q) = \inf_{q \in \mathbb{Q}, \, q > x} G(q) = G(x),$$

and analogously

$$\liminf_{k \to \infty} F_{h(k)}(x) \ge \sup_{q \in \mathbb{Q}, \, q < x} \liminf_{k \to \infty} F_{h(k)}(q) = \sup_{q \in \mathbb{Q}, \, q < x} G(q) = G(x).$$

Since (μ_h) is tight, we have also

$$\lim_{x \to -\infty} F_h(x) = 0, \qquad\qquad \lim_{x \to +\infty} F_h(x) = 1$$

uniformly with respect to h, hence $G(-\infty) = 0$ and $G(+\infty) = 1$. Notice now that the nondecreasing function

$$F(x) := \lim_{y \downarrow x} G(y)$$

is right continuous, and still satisfies $F(-\infty) = 0$ and $F(+\infty) = 1$, therefore (according to Theorem 6.23) F is the repartition function of a probability measure μ. Since $F = G$ on $\mathbb{R} \setminus E$, we have $F_{h(k)} \to F$ pointwise on $\mathbb{R} \setminus E$, and this proves the weak convergence of $\mu_{h(k)}$ to μ.

$\qquad\square$

The following theorem provides a characterization of the weak convergence in terms of convergence of the integrals of continuous and bounded functions.

Theorem 6.28. *Let μ_h, μ be probability measures in $\mathbb{R}$. Then μ_h weakly converge to μ if and only if*

$$\lim_{h \to \infty} \int_{\mathbb{R}} g \, d\mu_h = \int_{\mathbb{R}} g \, d\mu \qquad \forall g \in C_b(\mathbb{R}). \tag{6.32}$$

Proof. Assuming that $\mu_h \to \mu$ weakly, we denote by F_h and F the corresponding repartition functions and fix $g \in C_b(\mathbb{R})$. Let $M = \sup |g|$ and $\varepsilon > 0$. By Exercise 6.25 the sequence (μ_h) is tight, so that we can find $t > 0$ satisfying $\mu_h (\mathbb{R} \setminus (-t, t]) < \varepsilon$ for any $h \in \mathbb{N}$; we may assume (possibly choosing a larger t) that also $\mu (\mathbb{R} \setminus (-t, t]) < \varepsilon$ and that both $-t$ and t are points where the repartition functions are converging. Thanks to the uniform continuity of g in $[-t, t]$ we can find $\delta > 0$ such that

$$x, y \in [-t, t], \ |x - y| < \delta \qquad \Longrightarrow \qquad |g(x) - g(y)| < \varepsilon. \tag{6.33}$$

Hence, we can find points $t_1, \ldots, t_n$ in $[-t, t]$ such that $t_1 = -t$, $t_n = t$, there is convergence of the repartition functions in all points t_i, and $t_{i+1} - t_i < \delta$ for $i = 1, \ldots, n-1$. By (6.33) it follows that $\sup_{(-t,t]} |g - f| < \varepsilon$, where

$$f := \sum_{i=1}^{n-1} g(t_i) \mathbb{1}_{(t_i, t_{i+1}]}.$$

Splitting the integrals on $\mathbb{R}$ as the sum of an integral on $(-t, t]$ and an integral on $(-t, t]^c$ we have

$$\left| \int_{\mathbb{R}} g \, d\mu_h - \int_{(-t,t]} f \, d\mu_h \right| \leq M\varepsilon + \varepsilon = (M+1)\varepsilon \qquad \forall h \in \mathbb{N}, \quad (6.34)$$

and analogously

$$\left| \int_{\mathbb{R}} g \, d\mu - \int_{(-t,t]} f \, d\mu \right| \leq M\varepsilon + \varepsilon = (M+1)\varepsilon. \qquad (6.35)$$

Since

$$\int_{(-t,t]} f \, d\mu_h = \sum_{i=1}^{n-1} g(t_i) \left[F_h(t_{i+1}) - F_h(t_i) \right]$$

$$\to \sum_{i=1}^{n-1} g(t_i) \left[F(t_{i+1}) - F(t_i) \right] = \int_{(-t,t]} f \, d\mu,$$

adding and subtracting $\int_{(-t,t]} f \, d\mu_h$, and using (6.34) and (6.35), we conclude that

$$\limsup_{h \to \infty} \left| \int_{\mathbb{R}} g \, d\mu_h - \int_{\mathbb{R}} g \, d\mu \right| \leq (M+1)\varepsilon.$$

Since ε is arbitrary, (6.32) is proved.

Conversely, assume that (6.32) holds. Given $x \in \mathbb{R}$, define the open set $A = (-\infty, x)$; we can easily find $(g_k) \subseteq C_b(\mathbb{R})$ monotonically converging to $\mathbb{1}_A$ and deduce from (6.32) the inequality

$$\liminf_{h \to \infty} \mu_h(A) \geq \sup_{k \in \mathbb{N}} \liminf_{h \to \infty} \int_{\mathbb{R}} g_k \, d\mu_h = \sup_{k \in \mathbb{N}} \int_{\mathbb{R}} g_k \, d\mu = \mu(A).$$

Analogously, using a sequence $(g_k) \subseteq C_b(\mathbb{R})$ such that $g_k \downarrow \mathbb{1}_C$, with $C = (-\infty, x]$, we deduce from (6.32) the inequality

$$\limsup_{h \to \infty} \mu_h(C) \leq \inf_{k \in \mathbb{N}} \limsup_{h \to \infty} \int_{\mathbb{R}} g_k \, d\mu_h = \inf_{k \in \mathbb{N}} \int_{\mathbb{R}} g_k \, d\mu = \mu(C).$$

Therefore we have convergence of the repartition functions for any $x \in \mathbb{R}$ such that $\mu(A) = \mu(C)$, i.e. for any x that is not an atom of μ. We conclude thanks to Exercise 1.5. $\qquad \square$

Notice that in (6.32) there is no mention to the order structure of $\mathbb{R}$, and only the metric structure (i.e. the space $C_b(\mathbb{R})$) comes into play. In a general context, of probability measures on a metric space (X, d) endowed with the Borel σ–algebra $\mathscr{B}(X)$, we say that μ_h weakly converge to μ if

$$\lim_{h \to \infty} \int_X g \, d\mu_h = \int_X g \, d\mu \qquad \text{for any function } g \in C_b(X).$$

Exercises

6.21 Show that the probability measures

$$\mu_h := \frac{1}{h} \sum_{i=1}^{h} \delta_{\frac{i}{h}}$$

weakly converge to the probability measure $\mathbb{1}_{[0,1]}\mathscr{L}^1$.

6.22 Let $F_h : \mathbb{R} \to \mathbb{R}$ be nondecreasing functions pointwise converging to a nondecreasing function $F : \mathbb{R} \to \mathbb{R}$ on a dense set $D \subset \mathbb{R}$. Show that $F_h(x) \to F(x)$ at all points x where F is continuous.

6.23 Consider all atomic measures of the form

$$\sum_{i=-h^2}^{h^2} a_i \delta_{\frac{i}{h}},$$

where $h \in \mathbb{N}$ and $a_{-h}, \ldots, a_h \geq 0$. Show that for any finite Borel measure μ in $\mathbb{R}$ there exists a sequence of measures (μ_h) of the previous form that weakly converges to μ.

6.24 Show that a family $\{\mu_i\}_{i \in I}$ of probability measures in $\mathbb{R}$ is tight if and only if (6.31) holds.

6.25 Show that any sequence (μ_h) of probability measures weakly convergent to a probability measure is tight. *Hint:* if μ is the weak limit and $\varepsilon > 0$ is given, choose an integer $n \geq 1$ such that $\mu([1-n, n-1]) > 1 - \varepsilon$ and points $x \in (-n, 1-n)$ and $y \in (n-1, n)$ where the repartition functions of μ_h are converging to the repartition function of μ.

6.26 We want to extend what was shown in this section from the realm of probability measures to that of finite measures. Let (μ_h), μ be finite positive Borel measures on $\mathbb{R}$, and let F_h, F be their repartition functions. Consider the following implications:

(a) $\lim_h \int_{\mathbb{R}} g \, d\mu_h = \int_{\mathbb{R}} g \, d\mu \qquad \forall g \in C_b(\mathbb{R})$ (that is, (6.32));
(b) $\lim_h \int_{\mathbb{R}} g \, d\mu_h = \int_{\mathbb{R}} g \, d\mu \qquad \forall g \in C_c(\mathbb{R})$;
(c) F_h converge to F at all points where F is continuous;
(d) F_h converge to F on a dense subset of $\mathbb{R}$;
(e) $\lim_h \mu_h(\mathbb{R}) = \mu(\mathbb{R})$;
(f) (μ_h) is tight.

Find an example where (b) holds but (a), (c), (e) do not hold and prove the following implications: $a \Rightarrow b, e$, $a \Rightarrow c$, $d \Leftrightarrow c$, $b \wedge e \Rightarrow c$, $d \wedge e \Rightarrow f$, $d \wedge f \Rightarrow e$, $d \wedge f \Rightarrow a$. As a corollary, if (e) holds (as it happens in the case when all μ_h and μ are probability measures) we obtain that $a \Leftrightarrow b \Leftrightarrow c \Leftrightarrow d \Rightarrow f$.

6.8. Fourier transform

The Fourier transform is a basic tool in Pure and Applied Mathematics, Physics and Engineering. Here we just mention a few basic facts, focussing on the use of this transform in Measure Theory and Probability.

Definition 6.29 (Fourier transform of a function). Let $f \in L^1(\mathbb{R}, \mathbb{C})$. We set

$$\hat{f}(\xi) := \int_{\mathbb{R}} f(x) e^{-ix\xi} \, dx \qquad \forall \xi \in \mathbb{R}.$$

The function $\hat{f}$ is called Fourier transform of f.

Since the map $\xi \mapsto f(x) e^{-i\xi x}$ is continuous, and bounded by $|f(x)|$, the dominated convergence theorem gives that $\hat{f}(\xi)$ is continuous. The same upper bound also shows that $\hat{f}$ is bounded, and $\sup |\hat{f}| \leq \|f\|_1$. More generally, the following result holds:

Theorem 6.30. *Let $k \in \mathbb{N}$ be such that $\int_{\mathbb{R}} |x|^k |f|(x) \, dx < \infty$. Then $\hat{f} \in C^k(\mathbb{R}, \mathbb{C})$ and*

$$D^p \hat{f}(\xi) = (-i)^p \widehat{x^p f}(\xi) \qquad \forall p = 0, \ldots, k.$$

The proof of Theorem 6.30 is a straightforward consequence of the differentiation theorem for integrals depending on a parameter (in this case, the ξ variable, see the Appendix):

$$D_\xi^p \int_{\mathbb{R}} f(x) e^{-ix\xi} \, dx = \int_{\mathbb{R}} D_\xi^p \left(f(x) e^{-ix\xi} \right) \, dx$$

$$= (-i)^p \int_{\mathbb{R}} x^p f(x) e^{-ix\xi} \, dx.$$

According to the previous result, the Fourier transform allows to transform differentiations (in the ξ variable) into multiplications (in the x variable), thus allowing an algebraic solution of many linear differential equations.

In the sequel we need an explicit expression of the Fourier transform of a Gaussian function. For $\sigma > 0$, let

$$\rho_\sigma(x) := \frac{e^{-|x|^2/(2\sigma^2)}}{(2\pi\sigma^2)^{1/2}} \tag{6.36}$$

be the rescaled Gaussian functions, already considered in Example 6.25. Then

$$\int_{\mathbb{R}} \rho_\sigma(x) e^{-i\xi x} \, dx = e^{-\xi^2 \sigma^2/2} \qquad \forall \xi \in \mathbb{R}. \tag{6.37}$$

The proof of this identity is sketched in Exercise 6.27.

Remark 6.31. (Discrete Fourier transform) If $f : \mathbb{R} \to \mathbb{R}$ is a $2T$-periodic function, then we can write the Fourier series (corresponding, up to a linear change of variables, to those considered in Chapter 5 for 2π-periodic functions)

$$f = \sum_{n \in \mathbb{Z}} a_n e^{in\frac{\pi}{T}x}, \qquad \text{in } L^2((-T, T); \mathbb{C}), \qquad (6.38)$$

with

$$a_n = \frac{1}{2T} \int_{-T}^{T} f(x) e^{-in\frac{\pi}{T}x}\, dx, \qquad e^{in\frac{\pi}{T}x} = \cos n\frac{\pi}{T}x + i \sin n\frac{\pi}{T}x.$$

$$(6.39)$$

Remark 6.32. (Inverse Fourier transform) For $g \in L^1(\mathbb{R}, \mathbb{C})$ we define *inverse Fourier transform* of f the function

$$\tilde{g}(x) := \frac{1}{2\pi} \int_{\mathbb{R}} g(\xi) e^{ix\xi}\, d\xi \qquad x \in \mathbb{R}.$$

It can be shown (see for instance Chapter VI.1 in [7]) that the maps $f \mapsto \hat{f}$ and $g \mapsto \tilde{g}$ are each the inverse of the other in the so-called Schwarz space $\mathscr{S}(\mathbb{R}, \mathbb{C})$ of smooth and rapidly decreasing functions at infinity:

$$\mathscr{S}(\mathbb{R}, \mathbb{C}) := \left\{ f \in C^\infty(\mathbb{R}, \mathbb{C}) : \lim_{|x| \to \infty} |x|^k |D^i f|(x) = 0 \quad \forall k, i \in \mathbb{N} \right\}.$$

In particular we have

$$f(x) = 2\pi \left(\widetilde{\frac{\hat{f}}{2\pi}} \right)(x) = \int_{\mathbb{R}} a_\xi e^{ix\xi}\, d\xi \qquad \text{with}$$

$$a_\xi := \frac{1}{2\pi} \int_{\mathbb{R}} f(x) e^{-i\xi x}\, dx.$$

These formulas can be viewed as the continuous counterpart of the discrete Fourier transform (6.38), (6.39). In this sense, a_ξ are generalized Fourier coefficients, corresponding to the "frequency" ξ. The difference with Fourier series is that any frequency is allowed, not only the integer multiples $n\pi/T$ of a given one.

6.8.1. Fourier transform of a measure

In this section we are concerned in particular with the concept of Fourier transform of a measure.

Definition 6.33 (Fourier transform of a measure).   Let μ be a finite measure on $\mathbb{R}$. We set

$$\hat{\mu}(\xi) := \int_{\mathbb{R}} e^{-ix\xi}\, d\mu(x) \qquad\qquad \forall \xi \in \mathbb{R}.$$

The function $\hat{\mu} : \mathbb{R} \to \mathbb{C}$ is called *Fourier transform* of μ.

Notice that Definition 6.29 is consistent with Definition 6.33, because $\hat{\mu} = \hat{f}$ whenever $\mu = f\mathscr{L}^1$. Notice also that, by the dominated convergence theorem, the function $\hat{\mu}$ is continuous. Moreover $\hat{\mu}(0) = \mu(\mathbb{R})$ and, by estimating from above the modulus of the integral with the integral of the modulus (see also Exercise 6.29), we obtain that $|\hat{\mu}(\xi)| \leq \mu(\mathbb{R})$ for all $\xi \in \mathbb{R}$. Still using the differentiation theorems under the integral sign, one can check that for $k \in \mathbb{N}$ the following implications hold:

$$\int_{\mathbb{R}} |x|^k\, d\mu(x) < \infty \quad\implies\quad \hat{\mu} \in C^k(\mathbb{R}, \mathbb{C}) \text{ and} \tag{6.40}$$

$$D^p \hat{\mu}(\xi) = (-i)^p \widehat{x^p \mu}(\xi) \ \forall p = 0, \ldots, k.$$

Let us see other basic examples of Fourier transforms of probability measures:

Example 6.34. (1) If $\mu = \delta_{x_0}$ then $\hat{\mu}(\xi) = e^{-ix_0\xi}$.
(2) If $\mu = p\delta_1 + q\delta_0$ (with $p + q = 1$) is the *Bernoulli measure* with parameter p, then $\hat{\mu}(\xi) = q + pe^{-i\xi}$.
(3) If

$$\mu = \sum_{i=0}^{n} \binom{n}{i} p^i q^{n-i} \delta_i$$

is the *binomial measure* with parameters $n,\ p$ then

$$\hat{\mu}(\xi) = (q + pe^{-i\xi})^n \qquad\qquad \forall \xi \in \mathbb{R}.$$

(4) If $\mu = e^{-x}\mathbb{1}_{(0,\infty)}(x)\mathscr{L}^1$ is the *exponential measure*, then

$$\hat{\mu}(\xi) = \frac{1}{1 + i\xi} \qquad\qquad \forall \xi \in \mathbb{R}.$$

(5) If $\mu = (2a)^{-1}\mathbb{1}_{(-a,a)}\mathscr{L}^1$ is the *uniform measure* in $[-a, a]$, then

$$\hat{\mu}(\xi) = \frac{\sin(a\xi)}{a\xi} \qquad\qquad \forall \xi \in \mathbb{R} \setminus \{0\}.$$

(6) If $\mu = [\pi(1 + x^2)]^{-1}\mathscr{L}^1$ is the *Cauchy measure*, then [2]

$$\hat{\mu}(\xi) = e^{-|\xi|} \qquad\qquad \forall \xi \in \mathbb{R}.$$

---

[2] This computation can be done using the residue theorem in complex analysis

Theorem 6.35. *Any finite measure μ in $\mathbb{R}$ is uniquely determined by its Fourier transform $\hat{\mu}$.*

Proof. For $\sigma > 0$ we denote by ρ_σ the rescaled Gaussian functions in (6.36). According to Exercise 6.27 we have

$$e^{-z^2\sigma^2/2} = \int_{\mathbb{R}} \rho_\sigma(w)e^{-izw}\,dw.$$

Setting $z = (x - y)/\sigma^2$, dividing both sides by $(2\pi\sigma^2)^{1/2}$ we deduce that

$$\rho_\sigma(x - y) = \frac{1}{(2\pi\sigma^2)^{1/2}} \int_{\mathbb{R}} \rho_\sigma(w)e^{-iw(x-y)/\sigma^2}\,dw.$$

Using Fubini-Tonelli theorem we obtain

$$\int_{\mathbb{R}} \rho_\sigma(x - y)d\mu(x) = \int_{\mathbb{R}} \frac{1}{(2\pi\sigma^2)^{1/2}} \left(\int_{\mathbb{R}} \rho_\sigma(w)e^{-iw(x-y)/\sigma^2}\,dw \right) d\mu(x)$$

$$= \int_{\mathbb{R}} \frac{\rho_\sigma(w)}{(2\pi\sigma^2)^{1/2}} \hat{\mu}\left(\frac{w}{\sigma^2}\right) e^{iyw/\sigma^2}\,dy.$$

$$\tag{6.41}$$

As a consequence, the integrals $h_\sigma(y) = \int_{\mathbb{R}} \rho_\sigma(y-x)\,d\mu(x)$ are uniquely determined by $\hat{\mu}$. But, still using the Fubini-Tonelli theorem, one can check the identity

$$\int_{\mathbb{R}} \left(\int_{\mathbb{R}} g(y)\rho_\sigma(x - y)\,dy \right) d\mu(x) = \int_{\mathbb{R}} h_\sigma(y)g(y)\,dy \quad \forall g \in C_b(\mathbb{R}).$$

$$\tag{6.42}$$

Passing to the limit as $\sigma \downarrow 0$ and noticing that (by Example 6.25, that provides the weak convergence of $\rho_\sigma\lambda$ to δ_0 as $\sigma \downarrow 0$, or a direct verification)

$$\int_{\mathbb{R}} g(y)\rho_\sigma(x - y)\,dy = \int_{\mathbb{R}} g(x - z)\rho_\sigma(z)\,dz \quad \rightarrow \quad g(x) \qquad \forall x \in \mathbb{R}$$

from the dominated convergence theorem we obtain that all integrals $\int_{\mathbb{R}} g\,d\mu$, for $g \in C_b(\mathbb{R})$, are uniquely determined. Hence μ is uniquely determined by its Fourier transform. $\qquad\square$

Remark 6.36. It is also possible to show an *explicit* inversion formula for the Fourier transform. Indeed, (6.42) holds not only for continuous functions, but also for bounded Borel functions; choosing $a < b$ that are not atoms of μ and $g = \mathbb{1}_{(a,b)}$, we have that $\int_{\mathbb{R}} g(x)\rho_\sigma(x - y)\,dy \to g(x)$ for μ–a.e. x (precisely for $x \notin \{a, b\}$), so that (6.42) and (6.41) give

$$\mu((a, b)) = \lim_{\sigma\downarrow 0} \int_a^b h_\sigma(y)\,dy = \lim_{\sigma\downarrow 0} \int_a^b \int_{\mathbb{R}} \frac{e^{-w^2/2\sigma^2}}{2\pi\sigma^2} \hat{\mu}(\frac{w}{\sigma^2})e^{iyw/\sigma^2}\,dw\,dy.$$

The change of variables $w = t\sigma^2$ and Fubini theorem give

$$\mu((a, b)) = \lim_{\sigma \downarrow 0} \frac{1}{2\pi} \int_{\mathbb{R}} e^{-t^2\sigma^2/2} \hat{\mu}(t) \frac{e^{itb} - e^{ita}}{it} \, dt, \qquad (6.43)$$

for all points $a < b$ that are not atoms of μ.

According to Theorem 6.28 we have the implication:

$$\mu_h \to \mu \text{ weakly} \qquad \Longrightarrow \qquad \hat{\mu}_h \to \hat{\mu} \text{ pointwise in } \mathbb{R}. \qquad (6.44)$$

The following theorem, due to Lévy, gives essentially the converse implication, allowing to deduce the weak convergence from the convergence of the Fourier transforms.

Theorem 6.37 (Lévy). *Let (μ_h) be probability measures in $\mathbb{R}$. If $f_h = \hat{\mu}_h$ pointwise converge in $\mathbb{R}$ to some function f, and if f is continuous at 0, then $f = \hat{\mu}$ for some probability measure μ in $\mathbb{R}$ and $\mu_h \to \mu$ weakly.*

Proof. Let us show first that (μ_h) is tight. Fixed $a > 0$, taking into account that $\sin \xi$ is an odd function and using the Fubini theorem we get

$$\int_{-a}^{a} \hat{\sigma}(\xi) \, d\xi = \int_{-a}^{a} \int_{\mathbb{R}} e^{-ix\xi} \, d\sigma(x) d\xi = \int_{\mathbb{R}} \int_{-a}^{a} \cos(x\xi) \, d\xi d\sigma(x)$$

$$= \int_{\mathbb{R}} \frac{2}{x} \sin(ax) \, d\sigma(x)$$

for any probability measure σ. Hence, using the inequalities $|\sin t| \le |t|$ for all t and $|\sin t| \le |t|/2$ for $|t| \ge 2$, we get

$$\frac{1}{a} \int_{-a}^{a} \left(1 - \hat{\sigma}(\xi)\right) d\xi = 2 - 2 \int_{\mathbb{R}} \frac{\sin(ax)}{ax} \, d\sigma(x)$$

$$= 2 \int_{\mathbb{R}} \left(1 - \frac{\sin(ax)}{ax}\right) d\sigma(x) \qquad (6.45)$$

$$\ge \sigma \left(\mathbb{R} \setminus \left[-\frac{2}{a}, \frac{2}{a}\right]\right).$$

For $\varepsilon > 0$ we can find, by the continuity of f at 0, $a > 0$ such that

$$\int_{-a}^{a} (1 - f(\xi)) \, d\xi < \varepsilon a.$$

By the dominated convergence theorem we get $h_0 \in \mathbb{N}$ such that

$$\int_{-a}^{a} \left(1 - \hat{\mu}_h(\xi)\right) d\xi < \varepsilon a \qquad \forall h \ge h_0. \qquad (6.46)$$

As $a^{-1} \int_{-a}^{a} (1 - \hat{\mu}_h(\xi)) \, d\xi \to 0$ as $a \downarrow 0$ for any fixed h, we infer that we can find $b \in (0, a]$ such that (6.46) holds with b replacing a for all $h \in \mathbb{N}$. From (6.45) we get $\mu_h (\mathbb{R} \setminus [-n, n]) < \varepsilon$ for all $h \in \mathbb{N}$, as soon as $n > 2/b$.

Being the sequence tight, we can extract a subsequence $(\mu_{h(k)})$ weakly converging to a probability measure μ and deduce from (6.44) that $f = \hat{\mu}$. It remains to show that the whole sequence (μ_h) weakly converges to μ: if this is not the case there exist $\varepsilon > 0$, $g \in C_b(\mathbb{R})$ and a subsequence $h'(k)$ such that

$$\left| \int_{\mathbb{R}} g \, d\mu_{h'(k)} - \int_{\mathbb{R}} g \, d\mu \right| \geq \varepsilon \qquad \forall k \in \mathbb{N}.$$

But, possibly extracting one more subsequence, we can assume that $\mu_{h'(k)}$ weakly converge to a probability measure σ; in particular

$$\left| \int_{\mathbb{R}} g \, d\sigma - \int_{\mathbb{R}} g \, d\mu \right| \geq \varepsilon > 0. \tag{6.47}$$

As we are assuming that $f_h = \hat{\mu}_h$ converge pointwise to $f = \hat{\mu}$ we obtain that $\hat{\sigma} = \lim_k \hat{\mu}_{h'(k)} = \hat{\mu}$, hence $\hat{\mu} = \hat{\sigma}$. From Theorem 6.35 we obtain that $\mu = \sigma$, contradicting (6.47). $\qquad\square$

Notice that just pointwise convergences of the Fourier transforms is not enough to conclude the weak convergence, unless we know that the limit function is continuous: let us consider, for instance, the rescaled Gaussian kernels used in the proof of Theorem 6.35 and let us consider the behaviour of the Gaussian measures $\mu_\sigma = \rho_\sigma \mathscr{L}^1$ as $\sigma \uparrow \infty$; in this case, from Exercise 6.27 we infer that the Fourier transforms are pointwise converging in $\mathbb{R}$ to the discontinuous function equal to 1 at $\xi = 0$ and equal to 0 elsewhere. In this case we don't have weak convergence of the measures: we have, instead, the so-called phenomenon of dispersion of the whole mass at infinity

$$\lim_{\sigma \uparrow \infty} \mu_\sigma (\mathbb{R} \setminus [-n, n]) = \lim_{\sigma \uparrow \infty} \mu_1 \left(\mathbb{R} \setminus [-\frac{n}{\sigma}, \frac{n}{\sigma}] \right) = \mu_1 (\mathbb{R} \setminus \{0\}) = 1$$

$$\forall n \in \mathbb{N}$$

and the family of measures μ_σ is far from being tight as $\sigma \uparrow \infty$.

Exercises

6.27 Check the identity (6.37).

6.28 $\star$ Show that $\hat{\mu}$ is uniformly continuous in $\mathbb{R}$ for any finite measure μ.

6.29 Let μ be a probability measure in $\mathbb{R}$. Show that if $|\hat{\mu}|$ attains its maximum at $\xi_0 \neq 0$, then there exist $x_0 \in \mathbb{R}$ and $c_n \in [0, \infty)$ such that

$$\mu = \sum_{n \in \mathbb{Z}} c_n \delta_{x_n} \qquad \text{with} \qquad x_n = x_0 + \frac{2n\pi}{\xi_0}.$$

Use this fact to show that $|\hat{\mu}| \equiv 1$ in $\mathbb{R}$ if and only if μ is a Dirac mass.

Chapter 7
The fundamental theorem of the integral calculus

In this section we give a closer look at a classical theme, namely the fundamental theorem of the integral calculus, looking for optimal conditions on f ensuring the validity of the formula

$$f(x) - f(y) = \int_y^x f'(s)\,ds.$$

Notice indeed that in the classical theory of the Riemann integration there is a gap between the conditions imposed to give a meaning to the integral $\int_a^x g(s)\,ds$ (*i.e.* Riemann integrability of g) and those that ensure its differentiability as a function of x (for instance, typically one requires the continuity of g). We will see that this gap basically disappears in Lebesgue's theory, and that there is a precise characterization of the class of functions representable as $c + \int_a^x g(s)\,ds$ for a suitable (Lebesgue) integrable function g and for some constant c.

The following definition is due to Vitali.

Definition 7.1 (Absolutely continuous functions). Let $I \subset \mathbb{R}$ be an interval. We say that $f : I \to \mathbb{R}$ is absolutely continuous if for any $\varepsilon > 0$ there exists $\delta > 0$ for which the implication

$$\sum_{i=1}^n (b_i - a_i) < \delta \qquad \Longrightarrow \qquad \sum_{i=1}^n |f(b_i) - f(a_i)| < \varepsilon \qquad (7.1)$$

holds for any finite family $\{(a_i, b_i)\}_{1 \leq i \leq n}$ of pairwise disjoint intervals contained in I.

An absolutely continuous function is obviously uniformly continuous, but the converse is not true, see Example 7.7.

Let $f : [a, b] \to \mathbb{R}$ be absolutely continuous. For any $x \in [a, b]$ define

$$F(x) = \sup_{\sigma \in \Sigma_{a,x}} \sum_{i=1}^n |f(x_i) - f(x_{i-1})|,$$

where $\Sigma_{a,x}$ is the set of all decompositions $\sigma = \{a = x_0 < x_1 < \cdots < x_n = x\}$ of $[a, x]$. F is called the *total variation* of f. Let us check that F is finite: let $\delta > 0$ be satisfying the implication (7.1) with $\varepsilon = 1$ and let us estimate from above a sum in the definition of F. Without loss of generality we can assume that $|x_i - x_{i-1}| < \delta/2$ for all $i = 1, \ldots, n - 1$, possibly adding more points (which increases the sum). Then, we can split the sum in families of intervals with total length larger than $\delta/2$ and less than δ (just keep adding a new interval to a family if the total length does not exceed δ and notice that if it exceeds δ, the total length is at least $\delta/2$); the number of these families is less than $\frac{2}{\delta}(x - a)$ and, as a consequence, (7.1) gives

$$F(x) \le \frac{2}{\delta}(x - a) + 1.$$

We set

$$f^+(x) = \frac{1}{2}\,(F(x) + f(x)), \qquad f^-(x) = \frac{1}{2}\,(F(x) - f(x)),$$

so that

$$f(x) = f^+(x) - f^-(x), \quad F(x) = f^+(x) + f^-(x), \qquad x \in [a, b].$$

Lemma 7.2. *Let $f : [a, b] \to \mathbb{R}$ be absolutely continuous and let F be its total variation. Then F, f^+, f^- are nondecreasing and absolutely continuous.*

Proof. Let $x \in [a, b)$, $y \in (x, b]$ and $\sigma = \{a = x_0 < x_1 < \cdots < x_n = x\}$. Then we have

$$F(y) \ge |f(y) - f(x)| + \sum_{i=1}^{n} |f(x_i) - f(x_{i-1})|.$$

Taking the supremum over all $\sigma \in \Sigma_{a,x}$, yields

$$F(y) \ge |f(y) - f(x)| + F(x),$$

which implies that F, f^+, f^- are nondecreasing. It remains to show that F is absolutely continuous. Let $\varepsilon > 0$ and let $\delta = \delta(\varepsilon) > 0$ be such that the implication (7.1) holds for all finite families (a_i, b_i) of pairwise disjoint intervals with $\sum_i (b_i - a_i) < \delta$. Let now (a_i, b_i) be a family of disjoint intervals with $\sum_i (b_i - a_i) < \delta$ and let us prove that $\sum_i |F(b_i) - $

$F(a_i)| < 2\varepsilon$. For any $i = 1, \ldots, n$ we can find $\sigma_i = \{a_i = x_{0,i} < x_{1,i} < \cdots < x_{n_i,i} = b_i\}$ such that

$$F(b_i) - F(a_i) < \frac{\varepsilon}{n} + \sum_{k=1}^{n_i} |f(x_{k,i}) - f(x_{k-1,i})|, \qquad 1 \leq i \leq n. \quad (7.2)$$

Indeed, if $a = y_0 < y_1 < \cdots < y_{m_i} = b_i$ is a partition such that

$$F(b_i) < \frac{\varepsilon}{n} + \sum_{k=1}^{m_i} |f(y_k) - f(y_{k-1})|$$

we can assume with no loss of generality (adding one more element to the partition if necessary) that $y_k = a_i$ for some k; then, it suffices to estimate the first k terms of the above sum with $F(a_i)$, and to call $x_{0,i} = y_k, \ldots, x_{m_i-k+1,i} = y_{m_i}$ to obtain (7.2) with $n_i = m_i - k + 1$. Adding the inequalities (7.2) and taking into account that the union of the disjoint intervals $(x_{k,i-1}, x_{k,i})$ (for $1 \leq i \leq n, 0 \leq k \leq n_i$) has length less than δ, from the absolute continuity property of f we get

$$\sum_{i=1}^{n} (F(b_i) - F(a_i)) < \varepsilon + \varepsilon = 2\varepsilon.$$

This proves that F is absolutely continuous. $\qquad\square$

The absolute continuity property characterizes integral functions, as the following theorem shows.

Theorem 7.3. *Le $I = [a, b] \subset \mathbb{R}$. A function $f : I \to \mathbb{R}$ is representable as*

$$f(x) = f(a) + \int_a^x g(t)\, dt \qquad\qquad \forall x \in I \qquad (7.3)$$

for some $g \in L^1(I)$ if and only if f is absolutely continuous.

Proof. (Sufficiency) If f is representable as in (7.3), we have

$$|f(x) - f(y)| \leq \int_x^y |g(s)|\, ds \qquad\qquad \forall x, y \in I, \; x \leq y.$$

Hence, setting $A = \cup_i (a_i, b_i)$, the absolute continuity property follows by the implication

$$\mathscr{L}^1(A) < \delta \qquad \Longrightarrow \qquad \int_A |g|\, ds < \varepsilon.$$

The existence, given $\delta > 0$, of $\varepsilon > 0$ with this property is ensured by Exercise 6.14 (with $\mu = \mathscr{L}^1$ and $\nu = g\mathscr{L}^1$).

(Necessity) According to Lemma 7.2, we can write f as the difference of two nonincreasing absolutely continuous functions. Hence, we can assume with no loss of generality that f is nonincreasing, and possibly adding to f a constant we shall assume that $f(a) = 0$. We extend f to the whole of $\mathbb{R}$ setting $f \equiv 0$ in $(-\infty, a)$ and $f \equiv f(b)$ in (b, ∞). It is clear that this extension, that we still denote by f, retains the monotonicity and absolute continuity properties.

By Theorem 6.23 we obtain a unique finite measure ν on $(\mathbb{R}, \mathscr{B}(\mathbb{R}))$ without atoms (because f is continuous) such that f is the repartition function of ν. Since f is constant on $(-\infty, a)$ and on $(b, +\infty)$, we obtain that ν is concentrated on I, so that

$$f(x) = \nu\left((-\infty, x]\right) = \nu\left((a, x]\right) \qquad \forall x \in \mathbb{R}. \qquad (7.4)$$

Now, if we were able to show that $\nu \ll \mathbb{1}_I \mathscr{L}^1$, by the Radon–Nikodym theorem we would find $g \in L^1(I)$ such that $\nu = g\mathscr{L}^1$, so that (7.4) would give

$$f(x) = \int_a^x g(s)\, ds \qquad \forall x \in I.$$

Hence, it remains to show that $\nu \ll \mathbb{1}_I \mathscr{L}^1$. Taking into account the identity $\nu((a, b)) = f(b) - f(a)$, the absolute continuity property can be rewritten as follows: for any $\varepsilon > 0$ there exists $\delta > 0$ such that

$$\mathscr{L}^1(A) < \delta \qquad \Longrightarrow \qquad \nu(A) \leq \varepsilon$$

for any finite union of open intervals $A \subset I$. But, by approximation, the same implication holds for all open sets, because any such set is the countable union of open intervals. By Proposition 1.24, ensuring an approximation from above with open sets, the same implication holds for Borel sets $B \subset I$ as well. This proves that $\nu \ll \mathbb{1}_I \mathscr{L}^1$ and concludes the proof. $\qquad\qquad\square$

We will need the following nice and elementary covering theorem.

Theorem 7.4 (Vitali covering theorem). *Let $\{B_{r_i}(x_i)\}_{i \in I}$ be a finite family of balls in a metric space (X, d). Then there exists $J \subset I$ such that the balls $\{B_{r_i}(x_i)\}_{i \in J}$ are pairwise disjoint, and*

$$\bigcup_{i \in I} B_{r_i}(x_i) \subset \bigcup_{i \in J} B_{3r_i}(x_i). \qquad (7.5)$$

Proof. We proceed as follows: first we pick a ball with largest radius, then we remove all balls that intersect the first chosen ball and choose a second ball of largest radius among the remaining ones. We continue removing all balls that intersect the second chosen ball and picking a third ball of largest radius among the remaining ones, and so on. The process stops when either there is no ball left, *i.e.* when the remaining balls intersect at least one of the already chosen balls. The family of chosen balls is disjoint by construction. If $x \in B_{r_i}(x_i)$ and the ball $B_{r_i}(x_i)$ has not been chosen, then there is a chosen ball $B_{r_j}(x_j)$ intersecting it, so that $d(x_i, x_j) < r_i + r_j$. Moreover, if $B_{r_j}(x_j)$ is the *first* chosen ball with this property, then $r_j \geq r_i$ (otherwise, if $r_i > r_j$, either the ball $B_{r_i}(x_i)$ or a ball with larger radius would have been chosen, instead of $B_{r_j}(x_j)$), so that $d(x_i, x_j) < 2r_j$. It follows that

$$d(x, x_j) \leq d(x, x_i) + d(x_i, x_j) < r_i + 2r_j \leq 3r_j.$$

As x is arbitrary, this proves (7.5). $\square$

It is natural to think that the function g in (7.3) is, as in the classical fundamental theorem of integral calculus, the derivative of f. This is true, but far from being trivial, and it follows by the following weak continuity result (due to Lebesgue) of integrable functions. We state the result even in more then one variable, as the proof in this case does not require any extra difficulty.

Theorem 7.5 (Continuity in mean). *Let $f \in L^1(\mathbb{R}^n)$. Then, for $\mathscr{L}^n-$a.e. $x \in \mathbb{R}^n$ we have*

$$\lim_{r \downarrow 0} \frac{1}{\omega_n r^n} \int_{B_r(x)} |f(y) - f(x)|\, dy = 0.$$

The terminology "continuity in mean" can be explained as follows: it is easy to show that the integral means

$$\frac{1}{\omega_n r^n} \int_{B_r(x)} f(y)\, dy$$

of a continuous function f converge to $f(x)$ as $r \downarrow 0$ for any $x \in \mathbb{R}^n$, because they belong to the interval

$$[\min_{\overline{B}_r(x)} f, \max_{\overline{B}_r(x)} f].$$

The previous theorem tells us that the same convergence occurs, for $\mathscr{L}^n-$a.e. $x \in \mathbb{R}^n$, for *any* integrable function f. This simply follows by the inequality

$$\left| \frac{1}{\omega_n r^n} \int_{B_r(x)} f(y)\, dy - f(x) \right| = \frac{1}{\omega_n r^n} \left| \int_{B_r(x)} f(y) - f(x)\, dy \right|$$

$$\leq \frac{1}{\omega_n r^n} \int_{B_r(x)} |f(y) - f(x)|\, dy.$$

By the local nature of this statement, the same property holds for locally integrable functions.

Proof of Theorem 7.5. Given ε, $\delta > 0$ and an open ball $B = B_R(0)$, it suffices to check that the set

$$A := \left\{ x \in B : \limsup_{r \downarrow 0} \frac{1}{\omega_n r^n} \int_{B_r(x)} |f(y) - f(x)|\, dy > 2\varepsilon \right\}$$

has Lebesgue measure less than $(3^n + 1)\delta$. To this aim, we write f as the sum of a "good" part g and a "bad", but small, part h, *i.e.* $f = g + h$ with $g : B \to \mathbb{R}$ bounded and continuous, and $\|h\|_{L^1(B')} < \varepsilon\delta$, with $B' = B_{R+1}(0)$; this decomposition is possible, because Proposition 3.16 ensures the density of bounded continuous functions in $L^1(B)$.

The continuity of g gives

$$\lim_{r \downarrow 0} \frac{1}{\omega_n r^n} \int_{B_r(x)} |g(y) - g(x)|\, dy = 0 \qquad \forall x \in B.$$

Hence, as $f = g + h$, we have $A \subset A_1$, where

$$A_1 := \left\{ x \in B : \limsup_{r \downarrow 0} \frac{1}{\omega_n r^n} \int_{B_r(x)} |h(y) - h(x)|\, dy > 2\varepsilon \right\}.$$

Then, it suffices to show that $\mathscr{L}^n(A_1) \leq (3^n + 1)\delta$. By the triangle inequality, we have also $A_1 \subset A_2 \cup A_3$ with

$$A_2 := \{ x \in B : |h(x)| > \varepsilon \}$$

and

$$A_3 := \left\{ x \in B : \sup_{r \in (0,1)} \frac{1}{\omega_n r^n} \int_{B_r(x)} |h(y)|\, dy > \varepsilon \right\}.$$

Markov inequality ensures that $\mathscr{L}^n(A_2) \leq \|h\|_{L^1(B)}/\varepsilon < \delta$, so that we need only to show that $\mathscr{L}^n(A_3) \leq 3^n\delta$.

Since $x \mapsto \int_{B_r(x)} |h(y)|\, dy$ is continuous, we have that

$$x \mapsto \sup_{r \in (0,1)} \frac{1}{\omega_n r^n} \int_{B_r(x)} |h(y)|\, dy$$

is lower semi continuous, hence A_3 is open. Notice also that for any $x \in A_3$ there exists $r \in (0, 1)$, depending on x, such that

$$\int_{B_r(x)} |h(y)|\, dy > \varepsilon \omega_n r^n.$$

Let $K \subset A_3$ be a compact set and let $\{B(x_i, r_i)\}_{i \in I}$ be a finite family of these balls whose union covers K. By applying Vitali's covering theorem to this family of balls, we can find a disjoint subfamily $\{B_{r_i}(x_i)\}_{i \in J}$ such that the union of the enlarged balls $B_{3r_i}(x_i)$ still covers K. Adding the previous inequalities with $x = x_i$ and $r = r_i$ and summing in $i \in J$, since all balls $B_{r_i}(x_i)$ are contained in B' we get

$$\mathscr{L}^n(K) \le \sum_{i \in J} \omega_n (3 r_i)^n \le \frac{3^n}{\varepsilon} \sum_{i \in J} \int_{B_{r_i}(x_i)} |h(y)|\, dy$$

$$\le \frac{3^n}{\varepsilon} \int_{B'} |h(y)|\, dy \le 3^n \delta.$$

As K is arbitrary we obtain that $\mathscr{L}^n(A_3) \le 3^n \delta$. $\qquad\qquad\square$

Since the continuity in mean is a local property, it is not difficult to extend the previous result to locally integrable functions. By applying this extended theorem to a characteristic function $f = \mathbb{1}_E$ we get

$$\begin{cases} \displaystyle \lim_{r \downarrow 0} \frac{\mathscr{L}^n(E \cap B_r(x))}{\omega_n r^n} = 1 & \text{for } \mathscr{L}^n\text{--a.e. } x \in E \\[3mm] \displaystyle \lim_{r \downarrow 0} \frac{\mathscr{L}^n(E \cap B_r(x))}{\omega_n r^n} = 0 & \text{for } \mathscr{L}^n\text{--a.e. } x \in \mathbb{R}^n \setminus E \end{cases}$$

for any $E \in \mathscr{B}(\mathbb{R}^n)$; points of the first type are called *density points*, whereas points of the second type are called *rarefaction points*.

Using the continuity in mean of integrable functions we obtain the fundamental theorem of calculus within the (natural) class of absolutely continuous functions.

Theorem 7.6. *Let $I \subset \mathbb{R}$ be an interval and let $f : I \to \mathbb{R}$ be absolutely continuous. Then f is differentiable at $\mathscr{L}^1$–a.e. point of I. In addition f' is Lebesgue integrable in I and*

$$f(x) = f(a) + \int_a^x f'(s)\, ds \qquad \forall x \in I. \tag{7.6}$$

Proof. Let g be as in (7.3), let $x_0 \in I$ be a point where

$$\lim_{r \downarrow 0} \frac{1}{r} \int_{x_0 - r}^{x_0 + r} |g(s) - g(x_0)|\, ds = 0 \tag{7.7}$$

and notice that

$$\frac{f(x_0 + r) - f(x_0)}{r} = \frac{1}{r} \int_{x_0}^{x_0+r} g(s)\, ds$$

$$= g(x_0) + \frac{1}{r} \int_{x_0}^{x_0+r} g(s) - g(x_0)\, ds$$

for $r > 0$. Hence, passing to the limit as $r \downarrow 0$, from (7.7) we get $f'_+(x_0) = g(x_0)$; a similar argument shows that $f'_-(x_0) = g(x_0)$. As, according to the previous theorem, $\mathscr{L}^1$–a.e. point x_0 satisfies (7.7), we obtain that f is differentiable, with derivative equal to g, $\mathscr{L}^1$–a.e. in I. It suffices to replace g with f' in (7.3) to obtain (7.6). $\qquad\square$

One might think that differentiability $\mathscr{L}^1$–a.e. and integrability of the derivative are sufficient for the validity of (7.6) (these are the minimal requirements to give a meaning to the formula). However, this is not true, as the Heaviside function $\mathbb{1}_{(0,\infty)}$ fulfils these conditions but fails to be (absolutely) continuous. Then, one might think that one should require also the continuity of f to have (7.6). It turns out that not even this is enough: we build in the next example the *Cantor-Vitali function*, also called devil's staircase: a continuous function having derivative equal to 0 $\mathscr{L}^1$–a.e., but not constant. This example shows why a stronger condition, namely the *absolute* continuity, is needed.

Example 7.7 (Cantor–Vitali function). Let

$$X := \{f \in C([0, 1]) : \ f(0) = 0, \ f(1) = 1\}.$$

This is a closed subspace of the complete metric space $C([0, 1])$, hence X is complete as well. For any $f : [0, 1] \mapsto \mathbb{R}$ we set

$$Tf(x) := \begin{cases} f(3x)/2 & \text{if } 0 \leq 3x \leq 1, \\ 1/2 & \text{if } 1 < 3x < 2, \\ 1/2 + f(3x - 2)/2 & \text{if } 2 \leq 3x \leq 3. \end{cases} \qquad (7.8)$$

It is easy to see that T maps X into X, and that T is a contraction (with Lipschitz constant equal to 1/2). Hence, by the contraction principle, there is a unique $f \in X$ such that $Tf = f$.

Let us check that f has zero derivative $\mathscr{L}^1$–a.e. in $[0, 1]$. As $f = Tf$, f is constant, and equal to 1/2, in $(1/3, 2/3)$. Inserting this information again in the identity $f = Tf$ we obtain that f is locally constant (equal to 1/4 and to 3/4) on $(1/9, 2/9) \cup (7/9, 8/9)$. Continuing in this way, one finds that f is locally constant on the union of 2^{n-1} intervals, each of length 3^{-n}, $n \geq 1$. The complement $C = [0, 1] \setminus A$ of the union A

of these intervals is Cantor's middle third set (see also Exercise 1.8), and since

$$\mathscr{L}^1(A) = \sum_{n=1}^{\infty} \frac{2^{n-1}}{3^n} = \frac{1}{2} \sum_{n=1}^{\infty} \left(\frac{2}{3}\right)^n = 1$$

we know that $\mathscr{L}^1(C) = 0$. At any point of A the derivative of f is obviously 0.

In connection with the previous example, notice also that f maps A, a set of full Lebesgue measure in $[0, 1]$, into the countable set $\{2^{-n}\}_{n\geq 1}$. On the other hand, it maps C, a Lebesgue negligible set, into $[0, 1]$, a set with strictly positive Lebesgue measure.

Exercises

7.1 Let $H : \mathbb{R} \to \mathbb{R}$ be satisfying the Lipschitz condition

$$|H(x) - H(y)| \leq C|x - y| \qquad \forall x, \, y \in \mathbb{R}$$

and let $f : [a, b] \to \mathbb{R}$ be an absolutely continuous function. Show that $H \circ f$ is absolutely continuous in $[a, b]$.

7.2 $\star$ Let $E \subseteq \mathbb{R}$ be a Borel set and assume that any $t \in \mathbb{R}$ is either a point of density or a point of rarefaction of E. Show that either $\mathscr{L}^1(E) = 0$ or $\mathscr{L}^1(\mathbb{R} \setminus E) = 0$. (*Remark:* the same result is true in $\mathbb{R}^n$, but with a much harder proof, see [3], 4.5.11).

7.3[Lipschitz change of variables] $\star$ Let $f : I = [a, b] \to \mathbb{R}$ be absolutely continuous (resp. Lipschitz). Show that

$$\int_{f(a)}^{f(b)} \varphi(y)\, dy = \int_a^b \varphi(f(x)) f'(x)\, dx$$

for any bounded (resp. integrable) Borel function $\varphi : f(I) \to \mathbb{R}$.

7.4 Use the previous exercise to show that, for any Lipschitz function $f : \mathbb{R} \to \mathbb{R}$ and any $\mathscr{L}^1$–negligible set $N \in \mathscr{B}(\mathbb{R})$, the derivative f' vanishes $\mathscr{L}^1$–a.e. on $f^{-1}(N)$.

Chapter 8
Measurable transformations

In this chapter we study the classical problem of the change of variables in the integral from a new viewpoint. We will compute how the Lebesgue measure in $\mathbb{R}^n$ changes under a sufficiently regular transformation, generalizing what we have already seen for linear, or affine, maps. As a byproduct we obtain a quite general change of variables formula for integrals with respect to the Lebesgue measure.

8.1. Image measure

We are given two measurable spaces $(X, \mathscr{E})$ and $(Y, \mathscr{F})$, a measure μ on $(X, \mathscr{E})$ and a $(\mathscr{E}, \mathscr{F})$–measurable mapping $F: X \to Y$. We define a measure $F_{\#}\mu$ in $(Y, \mathscr{F})$ by setting

$$F_{\#}\mu(I) := \mu(F^{-1}(I)), \qquad I \in \mathscr{F}. \tag{8.1}$$

It is easy to see that $F_{\#}\mu$ is well defined, by the measurability assumption on F, and σ-additive on $\mathscr{F}$. $F_{\#}\mu$ is called the *image measure of μ by F*.

The following change of variable formula is simple, but of a basic importance.

Proposition 8.1. *Let $\varphi: Y \to [0, \infty]$ be a $\mathscr{F}$–measurable function. Then we have*

$$\int_X \varphi(F(x)) \, d\mu(x) = \int_Y \varphi(y) \, dF_{\#}\mu(y). \tag{8.2}$$

Proof. By monotone approximation it is enough to prove (8.2) when φ is a simple function. By linearity of both sides we need only to consider functions φ of the form $\varphi = \mathbb{1}_I$, where $I \in \mathscr{F}$. In this case we have $\varphi \circ F = \mathbb{1}_{F^{-1}(I)}$, hence (8.2) reduces to (8.1). $\qquad\square$

In the following example we discuss the relation between the change of variables formula (8.2), that even on the real line involves *no* derivative, and the classical one. The difference is due to the fact that in (8.2) we are not using the density of $F_{\#}\mu$ with respect to $\mathscr{L}^1$. It is precisely in this density that the derivative of F shows up.

Example 8.2. Let $F : \mathbb{R} \to \mathbb{R}$ be of class C^1 and such that $F'(t) > 0$ for all $t \in \mathbb{R}$. Let A be the image of F (an open interval, by the assumptions made on F) and let $\psi : A \to \mathbb{R}$ be continuous. Then for any interval $[a, b] \subset A$ the following elementary formula of change of variables holds (just put $y = F(x)$ in the right integral):

$$\int_{F^{-1}(a)}^{F^{-1}(b)} \psi(F(x)) \, dx = \int_a^b \psi(y) \frac{1}{F'(F^{-1}(y))} \, dy.$$

On the other hand, choosing $\varphi = \psi \mathbb{1}_I$ with $I = [a, b]$ in (8.2), we have

$$\int_{F^{-1}(a)}^{F^{-1}(b)} \psi(F(x)) \, dx = \int_a^b \psi(y) \, dF_\# \mathcal{L}^1.$$

Hence, comparing the two expressions, we find

$$\int_a^b \psi(y) \frac{1}{F'(F^{-1}(y))} \, dy = \int_a^b \psi(y) \, dF_\# \mathcal{L}^1. \tag{8.3}$$

Since a, b and ψ are arbitrary, (8.3) can be interpreted by saying that $F_\# \mathcal{L}^1 \ll \mathcal{L}^1$ and

$$F_\# \mathcal{L}^1 = \frac{1}{F' \circ F^{-1}} \mathcal{L}^1.$$

In the next section, we shall generalize this formula to $\mathbb{R}^n$, and even in one space dimension we will see that the assumption that $F' > 0$ *everywhere* can be weakened (see also Exercise 8.3).

8.2. Change of variables in multiple integrals

We consider here the measure space $(\mathbb{R}^n, \mathscr{B}(\mathbb{R}^n), \mathscr{L}^n)$, where $\mathscr{L}^n$ is the Lebesgue measure.

We recall a few basic facts from calculus with several variables: given an open set $U \subset \mathbb{R}^n$ and a mapping $F : U \to \mathbb{R}^n$, F is said to be *differentiable* at $x \in U$ if there exists a linear operator $DF(x) \in L(\mathbb{R}^n; \mathbb{R}^n)$ [1] such that

$$\lim_{|h| \to 0} \frac{|F(x + h) - F(x) - DF(x)h|}{|h|} = 0.$$

The operator $DF(x)$ if exists is unique, and is called the *differential of* F *at* x. If F is affine, *i.e.* $F(x) = Tx + a$ for some $T \in L(\mathbb{R}^n; \mathbb{R}^n)$ and $a \in \mathbb{R}^n$, we have $DF(x) = T$ for all $x \in U$.

---

[1] $L(\mathbb{R}^n; \mathbb{R}^m)$ is the Banach space of all linear mappings $T : \mathbb{R}^n \to \mathbb{R}^m$ endowed with the sup norm $\|T\| = \sup\{|Tx| : x \in \mathbb{R}^n, \ |x| = 1\}$

If F is differentiable at $x \in U$ we define the *Jacobian determinant* $J_F(x)$ of F at x by setting

$$J_F(x) = \det DF(x).$$

If F is differentiable at any $x \in U$ and if the mapping $DF: U \to L(\mathbb{R}^n; \mathbb{R}^n)$ is continuous, we say that F is of class C^1. If, in addition, F is bijective between U and an open domain A and F^{-1} is of class C^1 in A, we say that F is a C^1 diffeomorphism of U onto A. In this case we have that $DF(x)$ is invertible and

$$D(F^{-1})(F(x)) = (DF(x))^{-1} \qquad \forall x \in U.$$

Finally, by Proposition 6.10 we know that if $T \in L(\mathbb{R}^n; \mathbb{R}^n)$ we have

$$\mathscr{L}^n(T(E)) = |\det T| \, \mathscr{L}^n(E) \qquad \forall E \in \mathscr{B}(\mathbb{R}^n). \tag{8.4}$$

8.3. Image measure of $\mathscr{L}^n$ by a C^1 diffeomorphism

In this section we study how the Lebesgue measure changes under the action of a C^1 map F. The relevant quantity will be the function $|J_F|$, which really corresponds to the distorsion factor of the measure.

Let $U \subset \mathbb{R}^n$ be open. The *critical* set C_F of $F \in C^1(U; \mathbb{R}^n)$ is defined by

$$C_F := \{x \in U : \ J_F(x) = 0\}.$$

Lemma 8.3. *The image $F(C_F)$ of the critical set is Lebesgue negligible.*

Proof. Let $K \subset C_F$ be a compact set and $\varepsilon > 0$; for any $x \in K$ the set $DF(x)(\overline{B}_1(0))$ is Lebesgue negligible (because DF is singular at x, so that $DF(x)(\mathbb{R}^n)$ is contained in a $(n-1)$-dimensional subspace of $\mathbb{R}^n$), hence we can find $\delta = \delta(\varepsilon, x) > 0$ such that

$$\mathscr{L}^n \left(\{z \in \mathbb{R}^n : \ \mathrm{dist}\,(z - F(x), DF(x)(\overline{B}_1(0))) < \delta\} \right) < \varepsilon.$$

By a scaling argument we get

$$\mathscr{L}^n \left(\{z \in \mathbb{R}^n : \ \mathrm{dist}\,(z - F(x), DF(x)(\overline{B}_r(0))) < \delta r\} \right) < \varepsilon r^n \qquad \forall r > 0.$$

On the other hand, since $|F(y) - F(x) - DF(x)(y - x)| < \delta r$ in $B_r(x)$, provided r is small enough, we get

$$F(B_r(x)) \subset \{z \in \mathbb{R}^n : \ \mathrm{dist}\,(z - F(x), DF(x)(\overline{B}_r(0)) < \delta r\}.$$

It follows that $B_r(x) \subset U$ and $\mathscr{L}^n(F(B_r(x))) < \varepsilon r^n$ for $r > 0$ small enough, depending on x.

Since the family of balls $\{B_{r/3}(x)\}_{x \in K}$ covers the compact set K, we can find a finite family $\{B_{r_i/3}(x_i)\}_{i \in I}$ whose union still covers K and extract from it, thanks to Vitali's covering theorem, a subfamily $\{B_{r_i/3}(x_i)\}_{i \in J}$ made by pairwise disjoint balls such that the union of the enlarged balls $\{B_{r_i}(x_i)\}_{i \in J}$ covers K. In particular, covering $F(K)$ by the union of $F(B_{r_i}(x_i))$ for $i \in J$, we get

$$\mathscr{L}^n(F(K)) \le \sum_{i \in J} \varepsilon r_i^n = \frac{3^n \varepsilon}{\omega_n} \sum_{i \in J} \omega_n \left(\frac{r_i}{3}\right)^n \le \frac{3^n \varepsilon}{\omega_n} \mathscr{L}^n(U).$$

Letting $\varepsilon \downarrow 0$ we obtain that $\mathscr{L}^n(F(K)) = 0$. Since K is arbitrary, by approximation (recall that C_F, being a closed subset of U, can be written as the countable union of compact subsets of U) we obtain that $\mathscr{L}^n(F(C_F)) = 0$. $\qquad\square$

The following theorem provides a necessary and sufficient condition for the absolute continuity of $F_\#\mathscr{L}^n$ with respect to $\mathscr{L}^n$, assuming a C^1 regularity of F.

Theorem 8.4. *Let $U \subset \mathbb{R}^n$ be an open set and let $F : U \to \mathbb{R}^n$ be of class C^1, whose restriction to $U \setminus C_F$ is injective. Then:*

(i) $F_\#(\mathbb{1}_U \mathscr{L}^n)$ is absolutely continuous with respect to $\mathscr{L}^n$ if and only if C_F is Lebesgue negligible.
(ii) If $F_\#(\mathbb{1}_U \mathscr{L}^n) \ll \mathscr{L}^n$ we have

$$F_\#(\mathbb{1}_U \mathscr{L}^n) = \frac{1}{|J_F|(F^{-1})} \mathbb{1}_{F(U \setminus C_F)} \mathscr{L}^n. \tag{8.5}$$

Proof. (i) If $\mathscr{L}^n(C_F) > 0$, we have $F_\#(\mathbb{1}_U \mathscr{L}^n)(F(C_F)) \ge \mathscr{L}^n(C_F) > 0$ and $F_\#(\mathbb{1}_U \mathscr{L}^n)$ fails to be absolutely continuous with respect to $\mathscr{L}^n$, because we proved in Lemma 8.3 that $F(C_F)$ is Lebesgue negligible.

Let G be the inverse of the restriction of F to the open set $U \setminus C_F$. The local invertibility theorem ensures that the domain $A = F(U \setminus C_F)$ of G is an open set, that G is of class C^1 in A and that $DG(y) = (DF)^{-1}(G(y))$ for all $y \in A$. Let us assume now that C_F is Lebesgue negligible and let us show that $F^{-1}(E)$ is Lebesgue negligible whenever $E \subset F(U)$ is Lebesgue negligible. Since we already know that C_F is $\mathscr{L}^n$–negligible set, we can assume with no loss of generality that $E \subset A$ and show that $G(E)$ is Lebesgue negligible. Let A_M be the open sets

$$A_M := \{y \in A : \|DG(y)\| < M\}.$$

We will prove that

$$\mathscr{L}^n(G(K)) \le (3M)^n \mathscr{L}^n(K) \tag{8.6}$$

for any compact set $K \subset A_M$. So, $F_\# \mathcal{L}^n \leq (3M)^n \mathcal{L}^n$ on the compact sets of A_M and therefore on the Borel sets; in particular

$$\mathcal{L}^n(G(E \cap A_M)) \leq (3M)^n \mathcal{L}^n(E \cap A_M) = 0,$$

and letting $M \uparrow \infty$ we obtain that $\mathcal{L}^n(G(E)) = 0$, because $E \subset A$.

In order to show (8.6) we consider a bounded open set B contained in A_M and containing K, and the family of balls $B_r(y) \subset B$ with $y \in K$ and $r > 0$. For any of these balls the mean value theorem gives (with $t = t(y, z) \in (0, 1)$)

$$|G(z) - G(y)| = |DG((1-t)y + tz)(z-y)| \leq M|z-y| \qquad \forall z \in B_r(y),$$

therefore $G(B_r(y)) \subset B_{Mr}(G(y))$ for any of these balls. Since the family of balls $\{B_{r/3}(y)\}_{y \in F}$ covers K, we can find a finite family $\{B_{r_i/3}(y_i)\}_{i \in I}$ whose union still covers K and extract from it, thanks to Vitali's covering theorem, a subfamily $\{B_{r_i/3}(y_i)\}_{i \in J}$ made by pairwise disjoint balls such that the union of the enlarged balls $\{B_{r_i}(y_i)\}_{i \in J}$ covers K. In particular, by our choice of the radii of the balls, the family $\{B_{Mr_i}(G(y_i))\}_{i \in J}$ covers $G(K)$. We have then

$$\mathcal{L}^n(G(K)) \leq \sum_{i \in J} \omega_n (Mr_i)^n = (3M)^n \sum_{i \in J} \omega_n \left(\frac{r_i}{3}\right)^n \leq (3M)^n \mathcal{L}^n(B).$$

Letting $B \downarrow K$ we obtain (8.6).

Let us prove (ii). We denote by h the Radon–Nikodym derivative of $F_\#(\mathbb{1}_U \mathcal{L}^n)$ with respect to $\mathcal{L}^n$; by Theorem 7.5 we have that

$$h(y) = \lim_{r \downarrow 0} \frac{1}{\omega_n r^n} \int_{B_r(y)} h(z)\, dz = \lim_{r \downarrow 0} \frac{\mathcal{L}^n(G(B_r(y)))}{\omega_n r^n},$$

$$\text{for } \mathcal{L}^n\text{-a.e. } y \in A.$$

Taking into account that $F_\#(\mathbb{1}_U \mathcal{L}^n)$ is concentrated on A, and that $1/|J_F| \circ F^{-1} = |J_G|$, it remains to prove that for all $y_0 \in A$ we have

$$\lim_{r \downarrow 0} \frac{\mathcal{L}^n(G(B_r(y_0)))}{\omega_n r^n} = |J_G|(y_0). \tag{8.7}$$

For the sake of simplicity we only consider the case when $y_0 = 0$ and $G(0) = 0$ (this is not restrictive, up to a translation in the domain and in the codomain). We divide the rest of the proof in two steps.

Step 1. We assume in addition that $DG(0) = I$ and show that

$$\lim_{r \downarrow 0} \frac{\mathcal{L}^n(G(B_r(0)))}{\omega_n r^n} = 1, \tag{8.8}$$

which is equivalent to (8.7) in this case.

Since $DF(0) = DG(0) = I$ we have by the definition of derivative,

$$\lim_{|y|\to 0} \frac{|F(x) - x|}{|x|} = 0, \qquad \lim_{|y|\to 0} \frac{|G(y) - y|}{|y|} = 0$$

So, for any $\varepsilon \in (0, 1)$ there exists $\delta_\varepsilon > 0$ such that if $|x| < \delta_\varepsilon$ we have $x \in U \setminus C_F$ and $|F(x) - x| < \varepsilon|x|$ and if $|y| < \delta_\varepsilon$ we have $y \in F(U \setminus C_F)$ and $|G(y) - y| < \varepsilon|y|$. It follows that $r < \delta_\varepsilon$ implies

$$|F(x)| < r \quad \forall x \in B_{(1-\varepsilon)r}(0), \qquad |G(y)| < (1+\varepsilon)r \quad \forall y \in B_r(0). \tag{8.9}$$

In particular

$$B_{(1-\varepsilon)r}(0) \subset G(B_r(0)) \subset B_{(1+\varepsilon)r}(0) \qquad \forall r < \delta_\varepsilon. \tag{8.10}$$

Now, by (8.10) it follows that

$$(1-\varepsilon)^n \leq \frac{\mathscr{L}^n(G(B_r(0)))}{\omega_n r^n} \leq (1+\varepsilon)^n,$$

provided $r < \delta_\varepsilon$, and this proves that (8.8) holds.

Step 2. Set $T = DG(0)$ and $H(x) = T^{-1}G(x)$, so that $DH(0) = I$. Then we have $G(B_r(0)) = T(H(B_r(0)))$ and so, thanks to (8.4),

$$\mathscr{L}^n(G(B_r(0))) = \mathscr{L}^n(T(H(B_r(0)))) = |\det T|\,\mathscr{L}^n(H(B_r(0))),$$

which implies

$$\lim_{r\downarrow 0} \frac{\mathscr{L}^n(G(B_r(0)))}{\omega_n r^n} = |\det T|\lim_{r\downarrow 0} \frac{\mathscr{L}^n(H(B_r(0)))}{\omega_n r^n} = |\det T|.$$

The proof is complete. $\qquad\qquad\square$

Example 8.5 (Polar and spherical coordinates). Let us consider the polar coordinates

$$(\rho, \theta) \mapsto (\rho\cos\theta, \rho\sin\theta).$$

Here $U = (0, \infty) \times (0, 2\pi)$ and the critical set is empty, because the modulus of the Jacobian determinant is ρ.

In the case of the spherical coordinates

$$(\rho, \theta, \phi) \mapsto (\rho\cos\theta\sin\phi, \rho\sin\theta\sin\phi, \rho\cos\phi)$$

we have $U = (0, \infty) \times (0, 2\pi) \times (0, \pi)$ and the critical set is empty, because the modulus of the Jacobian determinant is $-\rho^2\sin\phi$.

Theorem 8.6 (Change of variables formula). *Let $U \subset \mathbb{R}^n$ be an open set and let $F : U \to \mathbb{R}^n$ of class C^1, injective on $U \setminus C_F$. Then*

$$\int_{F(U)} \varphi(y)\, dy = \int_U \varphi(F(x))|J_F|(x)\, dx \qquad (8.11)$$

for any Borel function $\varphi : F(U) \to [0, +\infty]$.

Proof. We first see that it is not restrictive to assume that $C_F = \varnothing$; indeed, $F(C_F)$ is Lebegue negligible and so images of points in C_F do not affect the left hand side, while obviously points in C_F do not affect the right hand side. So, possibly replacing U with $U \setminus C_F$, we can assume that $C_F = \varnothing$.

By (8.2) and (8.5) we have

$$\int_{F(U)} \frac{\psi(y)}{|J_F|(F^{-1}(y))}\, dy = \int_U \psi(F(x))\, dx.$$

for any nonnegative Borel function ψ. We conclude choosing $\psi(y) = \varphi(y)|J_F|(F^{-1}(y))$. $\qquad\square$

Exercises

8.1 Let $(X, \mathscr{F})$, $(Y, \mathscr{G})$ and $(Z, \mathscr{H})$ be measurable spaces and let $f : X \to Y$, $g : Y \to Z$ be measurable maps. Show that

$$g_\#(f_\#\mu) = (g \circ f)_\#\mu$$

for any measure μ in $(X, \mathscr{F})$.

8.2 Let $f : \{0, 1\}^{\mathbb{N}} \to [0, 1]$ be the map associating to a sequence $(a_i) \subset \{0, 1\}$ the real number $\sum_i a_i 2^{-i-1} \in [0, 1]$. Show that

$$f_\#\left(\bigtimes_{i=0}^{\infty}(\frac{1}{2}\delta_0 + \frac{1}{2}\delta_1)\right) = \mathbb{1}_{[0,1]}\mathscr{L}^1.$$

8.3 ★ Show the existence of a strictly increasing and C^1 function $F : \mathbb{R} \to \mathbb{R}$ such that $F_\#\mathscr{L}^1$ is not absolutely continuous with respect to $\mathscr{L}^1$.

8.4 ★ ★ Remove the injectivity assumption in Theorem 8.4, showing that

$$F_\#(\mathbb{1}_U\mathscr{L}^n) = \sum_{x \in F^{-1}(y) \setminus C_F} \frac{1}{|JF|(x)}\mathbb{1}_{F(U \setminus C_F)}\mathscr{L}^n.$$

for any C^1 function $F : U \to \mathbb{R}^n$ with Lebesgue negligible critical set.

Appendix A

A.1. Continuity and differentiability of functions depending on a parameter

In this section we consider the following problem: we are given a metric space (X, d) and a measure space $(Y, \mathscr{F}, \mu)$. Given $f : X \times Y \to \mathbb{R}$, we assume that for all $x \in X$ the function $f(x, \cdot)$ is μ–integrable, so that the function $F : X \to \mathbb{R}$ given by

$$F(x) := \int_Y f(x, y) \, d\mu(y) \qquad x \in X$$

is well defined. We would like to understand under which conditions F, an integral depending on the parameter x, is continuous. When X is an open subset of $\mathbb{R}^n$ endowed with the Euclidean distance, it is also natural to investigate the differentiability properties of F.

Theorem A.1 (Continuity of F). *Assume that $f(\cdot, y)$ is continuous in X for μ-almost all $y \in Y$ and that there exists $m \in L^1(Y, \mu)$ satisfying*

$$\sup_{x \in X} |f(x, y)| \le m(y) \qquad \text{for } \mu\text{–a.e. } y \in Y. \tag{A.1}$$

Then F is bounded and continuous in X.

Proof. It is clear that $|F(x)| \le \|m\|_1$ for all $x \in X$. Continuity is a simple consequence of the dominated convergence theorem: indeed, if $x_n \in X$ converge to x, then $f(x_n, y)$ converge to $f(x, y)$ for μ-almost every y and the convergence is dominated because of (A.1). It follows that $F(x_n) \to F(x)$. $\qquad \square$

A more expressive way to state the continuity of F is to say that limit and integral commute, namely

$$\lim_{h \to \infty} \int_Y f(x_h, y) \, d\mu(y) = \int_Y \lim_{h \to \infty} f(x_h, y) \, d\mu(y).$$

The following example shows that if no uniform upper bound is imposed on f, then continuity might fail:

Example A.2. Let $X = Y = \mathbb{R}$, $\mu = \mathscr{L}^1$ and

$$
f(x, y) := \begin{cases} |x|(1 - |y||x|) & \text{if } |y||x| < 1; \\ \\ 0 & \text{if } |y||x| \geq 1. \end{cases}
$$

Then $F(x) = 1$ for all $x \neq 0$, while $F(0) = 0$. In this case the smallest possible function satisfying (A.1) is $|y|^{-1}$ which is not integrable.

Next, we assume that X is an open set of $\mathbb{R}^n$ endowed with the Euclidean distance and we investigate the differentiability of F. Under suitable assumption, we can commute derivative and integral, namely

$$
\frac{\partial}{\partial x_i} \int_Y f(x, y)\, d\mu(y) = \int_Y \frac{\partial f}{\partial x_i}(x, y)\, d\mu(y) \qquad \forall x \in X, i = 1, \ldots, n.
$$
(A.2)

Theorem A.3 (Differentiability of F). *Assume that for μ-almost all $y \in Y$ the function $f(\cdot, y)$ is differentiable in X with a continuous gradient $\nabla_x f(x, y)$ and that, for any ball $B_r(x_0) \subset X$, there exists $m \in L^1(Y, \mu)$ satisfying*

$$
|f(x_0, y)| + \sup_{x \in B_r(x_0)} |\nabla_x f|(x, y) \leq m(y) \qquad \text{for } \mu\text{–a.e. } y \in Y. \quad \text{(A.3)}
$$

Then $F \in C^1(X)$ and (A.2) holds.

Proof. We fix $x_0 \in X$, $i \in \{1, \ldots, n\}$ and $x_i = x + t_i e_i$ with $t_i \neq 0$ and $t_i \to 0$. The mean value theorem, applied for any y such that $f(\cdot, y) \in C^1(X)$, gives $\theta_i(y) \in (0, 1)$ satisfying

$$
\frac{F(x_0 + t_i e_i) - F(x_0)}{t_i} = \int_Y \frac{\partial f}{\partial x_i}(x_0 + \theta_i(y) t_i e_i, y)\, d\mu(y)
$$

For i large enough (as soon as $|t_i| < r$) the functions of y inside the integral are dominated by the function m in (A.3), hence we can pass to the limit with the dominated convergence theorem to get (notice that the measurability of $\partial f / \partial x_i(x_0, \cdot)$ follows by the same limiting process)

$$
\frac{\partial F}{\partial x_i}(x_0) = \int_Y \frac{\partial f}{\partial x_i}(x_0, y)\, d\mu(y).
$$

Finally, continuity of partial derivatives of F is a consequence of the previous theorem. $\qquad\square$

Of course similar statements can be given for k-th order derivatives of F, provided $f(\cdot, y)$ is k times differentiable and, for any ball $B_r(x_0) \subset \mathbb{R}^n$ there exists $m \in L^1(Y, \mu)$ satisfying

$$|f(x_0, y)| + \sup_{x \in B_r(x_0)} \sup_{|p| \le k} |D^p f|(x, y) \le m(y) \qquad \text{for } \mu\text{–a.e. } y \in Y$$

(here $p = (p_1, \dots, p_n)$ and $|p| = p_1 + \cdots + p_n$). Under this assumption one obtains that

$$D^p \int_Y f(x, y)\, d\mu(y) = \int_Y D_x^p f(x, y)\, d\mu(y) \qquad \text{whenever } |p| \le k.$$

A.2. The dual space of continuous functions

In this section we want to characterize the space $(C(X))^*$, dual space of $C(X)$, with (X, d) compact metric space. Recall that $C(X)$ is a Banach space, when endowed with the sup norm, regardless of any assumption on (X, d). Some knowledge of the basic terminology of Banach spaces (dual space, dual norm) is needed for this section.

We start with some notation: we shall denote by $\mathscr{M}(X)$ the space of *signed* measures μ, i.e. the real-valued and σ-additive set functions μ, defined on $\mathscr{B}(X)$, of the form $\mu = \mu^+ - \mu^-$ with $\mu^\pm$ positive and finite Borel measures satisfying $\mu^+ \perp \mu^-$.

This orthogonality condition ensures uniqueness of the decomposition of μ, as we will see in a moment; existence, instead, is just a consequence of the σ-additivity (see Section 6.5), but we shall not use this fact in the sequel.

For $\mu \in \mathscr{M}(X)$ we denote $|\mu| = \mu^+ + \mu^-$ its total variation measure, as in Section 6.5, and set

$$\|\mu\| := |\mu|(X) = \mu^+(X) + \mu^-(X). \tag{A.4}$$

In the next proposition we show that the decomposition $\mu = \mu^+ - \mu^-$ is unique, so that (A.4) is well posed, and that $\mathscr{M}(X)$ is a normed space. The completeness of $\mathscr{M}(X)$ will be a consequence of Theorem A.6, since any dual space is complete.

Proposition A.4. *For any $\mu \in \mathscr{M}(X)$ the decomposition $\mu = \mu^+ - \mu^-$ is unique. In addition $\mathscr{M}(X)$, endowed with the norm (A.4), is a normed space.*

Proof. Assume that $\mu = \mu^+ - \mu^- = \tilde{\mu}^+ - \tilde{\mu}^-$, with orthogonal decompositions. Let A be a Borel set where μ^+ is concentrated, so that μ^- is concentrated on $X \setminus A$, and let $\tilde{A}$ be an analogous Borel set for $\tilde{\mu}^\pm$. Since

$\mu \geq 0$ (respectively $\mu \leq 0$) on the subsets of A (respectively of $X \setminus A$) and the same property holds for $\tilde{A}$, we obtain that μ (and therefore $\mu^{\pm}$ and $\tilde{\mu}^{\pm}$) vanishes on subsets of $A \setminus \tilde{A}$ and of $\tilde{A} \setminus A$. On the other hand, if $B \subset A \cap \tilde{A}$ we have $\mu^-(B) = \tilde{\mu}^-(B) = 0$ and

$$\mu^+(B) = \mu(B) = \tilde{\mu}^+(B).$$

Analogously, if $B \subset (X \setminus A) \cap (X \setminus \tilde{A})$ we have $\mu^+(B) = \tilde{\mu}^+(B) = 0$ and $\mu^-(B) = \tilde{\mu}^-(B)$. This proves that $\mu^{\pm} = \tilde{\mu}^{\pm}$.

Now, stability of $\mathscr{M}(X)$ under multiplication with real constants and 1-homogeneity of the norm are obvious. Let us prove stability under addition and subadditivity of the norm: if $\mu = \mu^+ - \mu^-$ and $\nu = \nu^+ - \nu^-$ we can write as before $\mu = f|\mu|$ and $\nu = g|\nu|$ with $f, g : X \to [-1, 1]$. Then, setting $\sigma = |\mu| + |\nu|$, the Radon–Nikodým theorem gives $|\mu| = a\sigma$ and $|\nu| = b\sigma$ for suitable $a, b : X \to [0, 1]$, so that

$$\mu + \nu = f|\mu| + g|\nu| = (fa + gb)\sigma$$

and we may take $(fa + gb)^{\pm}\sigma$ as positive and negative parts of $\mu + \nu$. We obtain also

$$\|\mu + \nu\| = \int_X |fa + gb| \, d\sigma \leq \int_X |a| + |b| \, d\sigma = \|\mu\| + \|\nu\|.$$

This completes the proof of the proposition. $\qquad\qquad\square$

We shall also denote by $\mathscr{A}(X)$ the collection of open subsets of X and use the following characterization of set functions defined on $\mathscr{A}(X)$ which are restrictions of σ-additive measures defined on the Borel σ-algebra.

Proposition A.5. *Let (X, d) be a compact metric space and let $\alpha : \mathscr{A}(X) \to [0, +\infty]$ be a nondecreasing set function satisfying $\alpha(\varnothing) = 0$ and:*

  (i) *(continuity) if $A_n \in \mathscr{A}(X)$, $n \in \mathbb{N}$, monotonically converge from below to A, then $\alpha(A_n) \uparrow \alpha(A)$;*

  (ii) *(subadditivity) $\alpha(A_1 \cup A_2) \leq \alpha(A_1) + \alpha(A_2)$ for all $A_1, A_2 \in \mathscr{A}(X)$;*

  (iii) *(additivity on disjoint sets) $\alpha(A_1 \cup A_2) = \alpha(A_1) + \alpha(A_2)$ whenever $A_1 \in \mathscr{A}(X)$ and $A_2 \in \mathscr{A}(X)$ are disjoint.*

Then

$$\tilde{\alpha}(B) := \inf\{\alpha(A) : A \in \mathscr{A}(X), \ A \supset B\} \qquad (A.5)$$

is a σ-additive extension of α to $\mathscr{B}(X)$.

Proof. Notice first that α is σ–subadditive on $\mathscr{A}(X)$: indeed, if $A \subset \cup_i A_i$ and B is an open set with compact closure in A, then B is contained in the union of finitely many A_i's, so that (ii) gives

$$\alpha(B) \le \sum_{i=1}^{\infty} \alpha(A_i).$$

Since B is arbitrary, (i) gives $\alpha(A) \le \sum_i \alpha(A_i)$.

Now, if we take (A.5) as the definition of $\tilde{\alpha}$ for all subsets of X, Proposition 1.16 gives that $\tilde{\alpha}$ extends α and is σ–subadditive. Then, Theorem 1.17 gives that α is σ–additive on the Borel σ–algebra, provided we are able to show that any Borel set is $\tilde{\alpha}$–additive. Since the class of additive sets is a σ–algebra, suffices to show that any closed set is $\tilde{\alpha}$–additive.

To this aim, we first show that $\tilde{\alpha}$ is additive on distant sets, namely (recall that $\mathrm{dist}(U, V)$ is the infimum of the distances $d(x, y)$ for $x \in U$ and $y \in V$)

$$\tilde{\alpha}(B_1 \cup B_2) = \tilde{\alpha}(B_1) + \tilde{\alpha}(B_2) \qquad \text{whenever } \mathrm{dist}(B_1, B_2) > 0. \quad \text{(A.6)}$$

Indeed, if $A \supset B_1 \cup B_2$ is open we can consider the disjoint open sets

$$A_1 := \{x \in A : \ \mathrm{dist}(x, B_1) < \mathrm{dist}(x, B_2)\},$$

$$A_2 := \{x \in A : \ \mathrm{dist}(x, B_2) < \mathrm{dist}(x, B_2)\}$$

containing B_1 and B_2 respectively to get

$$\alpha(A) \ge \alpha(A_1 \cup A_2) = \alpha(A_1) + \alpha(A_2) \ge \tilde{\alpha}(B_1) + \tilde{\alpha}(B_2).$$

Since A is arbitrary the inequality $\ge$ in (A.6) follows, while the converse one is a consequence of subadditivity.

Let $F \subset X$ be closed, $B \subset X$ and let us prove that $\tilde{\alpha}(B \cap F) + \tilde{\alpha}(B \setminus F) \le \tilde{\alpha}(B)$ (the opposite inequality follows by subadditivity). Assuming with no loss of generality $\tilde{\alpha}(B) < \infty$ and setting

$$B_h := \left\{x \in B : \ 2^h > \mathrm{dist}(x, F) \ge 2^{h-1}\right\} \qquad h \in \mathbb{Z}$$

the additivity on distant sets gives

$$\sum_{h \in \mathbb{Z}} \tilde{\alpha}(B_{2h}) \le \tilde{\alpha}(B) < \infty, \qquad \sum_{h \in \mathbb{Z}} \tilde{\alpha}(B_{2h+1}) \le \tilde{\alpha}(B) < \infty$$

because all finite sums are made on distant sets, all contained in B. We have then that $\sum_{h\in\mathbb{Z}}\tilde{\alpha}(B_h)$ is convergent and, since the sets B_h are a partition of $B \setminus F$, using once more the additivity on distant sets we get

$$\tilde{\alpha}(B \cap F) + \tilde{\alpha}(B \setminus F) \leq \tilde{\alpha}(B \cap F) + \tilde{\alpha}\left(\bigcup_{h=-\infty}^{N} B_h\right) + \sum_{h=N+1}^{\infty} \tilde{\alpha}(B_h)$$

$$= \tilde{\alpha}\left((B \cap F) \cup \bigcup_{h=-\infty}^{N} B_h\right) + \sum_{h=N+1}^{\infty} \tilde{\alpha}(B_h)$$

$$\leq \tilde{\alpha}(B) + \sum_{h=N+1}^{\infty} \tilde{\alpha}(B_h)$$

for any $N \geq 1$. Letting $N \to \infty$ the inequality follows. $\square$

For $g \in C(X)$ we can define

$$\int_X g \, d\mu := \int_X g \, d\mu^+ - \int_X g \, d\mu^-.$$

In this way $\int g \, d\mu$ is linear w.r.t. g; in addition, since

$$\int_X h \, \mu = \int_0^\infty \mu^+(\{h > t\}) \, dt - \int_0^\infty \mu^-(\{h > t\}) \, dt$$

$$= \int_0^\infty \mu(\{h > t\}) \, dt$$

whenever h is nonnegative, splitting g in positive and negative part we obtain that $\int_X g \, d\mu$ is also linear w.r.t. to μ. Since

$$\left|\int_X g \, d\mu\right| \leq \int_X |g| \, d\mu^+ + \int_X |g| \, d\mu^- \leq \max|g|\|\mu\| = \|g\|\|\mu\|$$

$$\forall g \in C(X)$$

the functional

$$L_\mu(g) := \int_X g \, d\mu \qquad g \in C(X) \tag{A.7}$$

belongs to $(C(X))^*$ and satisfies $\|L_\mu\| \leq \|\mu\|$. The remarkable fact is that any element in the dual is representable in this form, and that equality holds. This will also prove that $\mathscr{M}(X)$ is a Banach space (with the definition of $\mathscr{M}(X)$ given above, independent of Section 6.5, it is not even totally obvious that it is a linear space!).

Theorem A.6 (Riesz). *Let (X, d) be a compact metric space. The space $(C(X))^*$ is, via (A.7), isomorphic and isometric to $\mathscr{M}(K)$. That is: all functionals L_μ belong to $(C(X))^*$ and, for any $L \in (C(X))^*$, there exists a unique $\mu \in \mathscr{M}(K)$ satisfying $L = L_\mu$. Finally, $\|L_\mu\| = \|\mu\|$.*

Proof. The proof will be achieved in three steps. In the first one we build an auxiliary positive finite measure μ^* and prove in the second one that μ^* provides the desired representation of L when L is nondecreasing. In the last one we achieve the general case and provide equality of the norms.

Step 1. Let $\alpha^* : \mathscr{A}(X) \to [0, +\infty)$ be defined by

$$\alpha^*(A) := \sup \{|L(g)| : \ |g| \le 1, \ \operatorname{supp} g \subset A\}.$$

Notice that $\alpha^*(X) \le \|L\|$ and that $\alpha^*(\varnothing) = 0$. Notice also that we can equivalently replace $|L(g)|$ with $L(g)$ inside the supremum and that a simple approximation argument gives

$$\alpha^*(A) \ge |L(g)| \qquad \text{whenever } |g| \le \mathbb{1}_A. \tag{A.8}$$

Indeed, if $|g| \le \mathbb{1}_A$ we can find continuous functions $g_n : X \to [-1, 1]$ convergent to g and with support contained in A. In addition, if L is monotone we have also

$$\alpha^*(A) \le L(\chi) \qquad \text{whenever } \mathbb{1}_A \le \chi. \tag{A.9}$$

We claim that α^* satisfies all the assumption of Proposition A.5. Indeed, if $g \in C(X)$ has support contained in A, since the support is compact we have $K \subset A_i$ for i large enough; it follows that $L(g) \le \alpha^*(A_i) \le \sup_j \alpha^*(A_j)$ and since g is arbitrary the continuity follows. In order to prove the subadditivity, given a continuous $g : X \to [-1, 1]$ with support K contained in $A_1 \cap A_2$, we can consider the disjoint compact sets $K \setminus A_1$ and $K \setminus A_2$ and a continuous function $\chi : X \to [0, 1]$ identically equal to 1 in a neighbourhood of $K \setminus A_1$ and identically equal to 0 in a neighbourhood of $K \setminus A_2$. It follows that $(1 - \chi)g$ has support contained in A_1 and χg has support contained in A_2, hence

$$L(g) = L((1 - \chi)g) + L(\chi g) \le \alpha^*(A_1) + \alpha^*(A_2).$$

Since g is arbitrary, the subadditivity of α^* follows. Finally, to prove the additivity on disjoint sets it suffices to notice that, given g_i with support in A_i and $|g_i| \le 1$, the function $g = g_1 + g_2$ has support in $A_1 \cup A_2$ and satisfies $L(g) = L(g_1) + L(g_2)$ and $|g| \le 1$.

By Proposition A.5 we obtain that α^* is the restriction to $\mathscr{A}(X)$ of a positive measure μ^*. Notice also that μ^* is finite, since

$$\mu^*(X) = \alpha^*(X) = \|L\|. \tag{A.10}$$

Step 2. Now we claim that $L_{\mu^*} \geq |L|$, namely $L_{\mu^*}(g) \geq |L(g)|$ for any nonnegative $g \in C(X)$. Also, we shall prove that if L is nondecreasing, namely $L(g) \geq 0$ whenever $g \in C(X)$ is nonnegative, then L_{μ^*} coincides with L. This proves already Riesz theorem for positive functionals.

By homogeneity, in the proof of the inequality $L_{\mu^*}(g) \geq |L(g)|$, it is not restrictive to assume $0 \leq g \leq 1$. Given an integer $N \geq 1$, let us consider the open sets $A_i := \{g > i/N\}$, $i = 0, \ldots, N-1$, and notice that

$$\frac{1}{N} + \sum_{i=1}^{N-1} \frac{1}{N} \mathbb{1}_{A_i} \geq g \geq \sum_{i=1}^{N-1} \mathbb{1}_{A_i}. \tag{A.11}$$

Now, given continuous functions $\chi_i : X \to [0, 1]$ satisfying $\mathbb{1}_{A_i} \leq \chi_i \leq \mathbb{1}_{A_{i-1}}$, $i = 1, \ldots, N$, we can use (A.8) to estimate

$$L_{\mu^*}(g) \geq \sum_{i=1}^{N-1} \frac{1}{N} \mu^*(A_i) \geq \sum_{i=1}^{N-1} \frac{1}{N} |L(\chi_{i+1})| \geq \left| L\left(\frac{1}{N} \sum_{i=2}^{N} \chi_i \right) \right|.$$

But, since

$$\frac{1}{N} + \sum_{i=1}^{N-1} \frac{1}{N} \chi_i \geq g \geq \sum_{i=1}^{N-1} \frac{1}{N} \chi_{i+1} \tag{A.12}$$

we can let $N \to \infty$ and use the continuity of L to get $L_{\mu^*}(g) \geq |L(g)|$.

If L is also monotone we can use the inequality (A.9) to get

$$L_{\mu^*}(g) - \frac{1}{N} \leq \sum_{i=1}^{N-1} \frac{1}{N} \mu^*(A_i) \leq \sum_{i=1}^{N-1} \frac{1}{N} L(\chi_i) = L\left(\frac{1}{N} \sum_{i=1}^{N-1} \chi_i \right).$$

Again we can let $N \to \infty$ and use (A.12) to get $L_{\mu^*}(g) = L(g)$.

Step 3. Now we define linear continuous functionals $L^\pm : C(X) \to \mathbb{R}$ by

$$L^+(g) := \frac{L_{\mu^*}(g) + L(g)}{2}, \qquad L^-(g) := \frac{L_{\mu^*}(g) - L(g)}{2}.$$

We have $L^+ + L^- = L_{\mu^*}$ and $L^+ - L^- = L$. In addition, by Step 2, $L^\pm$ are monotone.

Now we can apply the construction of Step 1 and use monotonicity in Step 2 to find positive finite measures $\mu^\pm$ such that $L^\pm = L_{\mu^\pm}$. It follows that

$$L = L^+ - L^- = L_{\mu^+} - L_{\mu^-} = L_\mu$$

and the representation of L follows. Analogously, we obtain that

$$L_{\mu^*} = L^+ + L^- = L_{\mu^+} + L_{\mu^-} = L_{\mu^+ + \mu^-}$$

so that $\mu^* = \mu^+ + \mu^-$. To conclude, we identify $\|\mu\|$ with $\|L\|$ and show that μ^+ and μ^- are orthogonal. The bound on $\|\mu\|$ follows by (A.10):

$$\|\mu\| = \mu^+(X) + \mu^-(X) = L_{\mu^+}(1) + L_{\mu^-}(1) = L_{\mu^*}(1) = \|L\|.$$

In order to show that $\mu^+ \perp \mu^-$, write $\mu^\pm = a^\pm \mu^*$ and use the identity $\mu^* = \mu^+ + \mu^-$ to get $a^+ + a^- = 1$ μ^*–a.e. in X. On the other hand the density of $C(X)$ in $L^2(X, \mu^*)$ and a truncation argument provide a sequence of continuous functions $g_n : X \to [-1, 1]$ convergent in $L^2(X, \mu^*)$ to the sign of $a^+ - a^-$, so that

$$\|L\| = \sup_{|g| \leq 1} |L_\mu(g)| = \sup_{|g| \leq 1} \left| \int (a^+ - a^-) g \, d\mu^* \right| = \int_X |a^+ - a^-| \, d\mu^*.$$

Hence

$$\int_X (1 - |a^+ - a^-|) \, d\mu^* = \mu^*(X) - \|L\| \leq 0.$$

Since $|a^+ - a^-| \leq 1$ it must be $|a^+ - a^-| = 1$ μ^*–a.e. in X. Since $a^\pm \in [0, 1]$ μ–a.e., this can only happen if $a^+ a^- = 0$ μ^*–a.e. in X, which means that μ^+ is orthogonal to μ^-. $\qquad\qquad \square$

Remark A.7. A similar result holds, with minor changes in the proof, if (X, d) is locally compact and separable, namely there exists an non-decreasing sequence of open sets with compact closure whose union is the whole of X. In this case $C(X)$ has to be replaced by $C_0(X)$, namely the closure in $C(X)$ of the space $C_c(X)$ of compactly supported functions, while $\mathcal{M}(X)$ remains unchanged.

Solutions of some exercises

In this chapter we provide solutions to the main exercises proposed in the text, and in particular of those marked with one or two $\star$.

Chapter 1

Exercise 1.1. All verifications are very simple and we omit them.

Exercise 1.2. We prove the statement for the translations, the proof for the dilations being similar. Fix $h \in \mathbb{R}$ and consider the class

$$\mathscr{F} := \{A \in \mathscr{B}(\mathbb{R}) \ : \ A + h \in \mathscr{B}(\mathbb{R})\}.$$

Then $\mathscr{F}$ is a σ–algebra containing the intervals, because the class $\mathscr{I}$ of intervals is invariant under translations. Therefore $\mathscr{F} \supset \sigma(\mathscr{I}) = \mathscr{B}(\mathbb{R})$. This proves that $A + h$ is Borel whenever A is Borel.

Exercise 1.3. Set $X = \mathbb{N}$ and $\mu := \sum_n \delta_n$. Then the sets $A_n := \{n, n + 1, \ldots\}$ satisfy $\mu(A_n) = +\infty$, but their intersection is empty.

Exercise 1.4. Let $A_n \uparrow A$ with $A_n, \ A \in \mathscr{A}$. Then the sets $B_n := A \setminus A_n$ satisfy $B_n \downarrow \varnothing$, so that by assumption $\mu(B_n) \downarrow \mu(\varnothing) = 0$. Since μ is finite, $\mu(B_n) = \mu(A) - \mu(A_n)$, so that $\mu(A_n) \uparrow \mu(A)$.

Exercise 1.5. For any $n \in \mathbb{N}^*$ the set A_n of all atoms x such that $\mu(\{x\}) \geq 1/n$ has at most cardinality $n\mu(X)$: indeed, if we choose k elements $x_1, \ldots, x_k$ in this sets, adding the inequalities $\mu(\{x_i\}) \geq 1/n$ we find $k/n \leq \mu(X)$, whence the upper bound on the cardinality of A_n follows.

    If μ is σ–finite, we choose $X_i \uparrow X$ with $X_i \in \mathscr{E}$ and $\mu(X_i) < \infty$ and repeat the previous argument with the sets $A_{i,n} := \{x \in A_\mu \cap X_i \ : \ \mu(\{x\}) \geq 1/n\}$, whose union gives A_μ. If not finiteness assumption is made, the statement fails: take $X = \mathbb{R}$, $\mathscr{E} = \mathscr{P}(\mathbb{R})$ and $\mu(A) = 0$ if $A = \varnothing$ and $\mu(A) = +\infty$ otherwise.

Exercise 1.6. Let μ be diffuse. First we prove that for all $\tau \in (0, 1)$ and all $A \in \mathscr{E}$ there exists a subset $B \in \mathscr{E}$ with $0 < \mu(B) < \tau\mu(A)$. Indeed, if this property fails for some τ and A, for all subsets B either $\mu(B) = 0$ or $\mu(B) \geq \tau\mu(A)$. Now, choose $B_1 \subset A$ with $\mu(B_1) \in (0, \mu(A))$ (this is possible by assumption), then $B_2 \subset A \setminus B_1$ with $\mu(B_2) \in (0, \mu(B_1))$ and so on. Since all these sets are contained in A, we have $\mu(B_i) \geq \tau\mu(A)$, and this contradicts the fact that they are disjoint.

Now, given $t \in (0, \mu(X))$ we define a sequence of pairwise disjoint sets B_i and numbers s_i as follows: first set

$$s_1 := \sup \{\mu(B) : \ \mu(B) \leq t\}$$

and then choose B_1 with $t \geq \mu(B_1) > s_1/2$; then recursively set

$$s_{n+1} := \sup \left\{\mu(B) : \ B \subset B_n^c, \ \mu(B) \leq t - \mu(B_n)\right\}$$

and choose $B_{n+1} \subset B_n^c$ with $t - \mu(B_n) \geq \mu(B_{n+1}) > s_{n+1}/2$. We now claim that $\mu(\cup_i B_i) = t$. If this property fails, then $\sum_i \mu(B_i) < t$ and the convergence of the series implies that $s_i \to 0$. On the other hand

$$s_i \geq \sup \left\{\mu(B) : \ B \subset X \setminus \bigcup_i B_i, \ \mu(B) \leq t - \sum_i \mu(B_i)\right\}$$

The previous property with $A = X \setminus \cup_i B_i$ and $\tau = (t - \sum_i \mu(B_i))/\mu(A)$ shows that the supremum in the right hand side (independent of i) is positive, contradicting the fact that $s_i \to 0$.

Exercise 1.7. Let X be a separable metric space and let $\mathscr{E} = \mathscr{B}(X)$. If $\mu(\{x\}) > 0$ for some $x \in X$, obviously μ is not diffuse. Conversely, if $A \in \mathscr{B}(X)$ is given, with $\mu(A) > 0$ and $\mu(B) \in \{0, \mu(A)\}$ for all $B \subset A$, we can fix a countable dense set $(x_i) \subset X$ and define

$$r_0 := \sup \left\{r \geq 0 \ : \ \mu(A \cap \overline{B}_r(x_0)) = 0\right\}.$$

Since $r \mapsto \mu(A \cap \overline{B}_r(x_0))$ is right continuous, the maximality of r_0 easily implies that $\mu(A \cap \overline{B}_{r_0}(x_0)) > 0$, and therefore $\mu(A \cap \overline{B}_{r_0}(x_0)) = \mu(A)$. Now we iterate this construction, setting $A_1 := A \cap \overline{B}_{r_0}(x_0)$, defining

$$r_1 := \sup \left\{r \geq 0 \ : \ \mu(A_1 \cap \overline{B}_r(x_1)) = 0\right\},$$

so that $\mu(A_1 \cap \overline{B}_{r_1}(x_1)) = \mu(A_1) = \mu(A)$. Continuing in this way, we have a nonincreasing family of sets (A_i) with $\mu(A_i) = \mu(A)$; it follows that $\mu(\cap_i A_i) = \mu(A) > 0$. On the other hand, any point $x \in \cap_i A_i$ satisfies

$$d(x, x_i) = r_i \qquad \forall i \in \mathbb{N}.$$

By the density of the family (x_i), this intersection contains at most one point (and at least one, because the measure is positive). It follows that this point is an atom of μ.

Exercise 1.8. Cantor's middle third set can be obtained as follows: let $C_0 = [0, 1]$, let C_1 the set obtained from C_0 by removing the interval $(1/3, 2/3)$, let C_2 be the set obtained from C_1 by removing the intervals $(1/9, 2/9)$ and $(7/9, 8/9)$, and so on. Each set C_n consists of 2^n disjoint closed intervals with length 3^{-n}, so that $\lambda(C_n) = (2/3)^n \to 0$. If follows that the intersection C of all sets C_n is a closed and λ–negligible set.

In order to show that C has the cardinality of continuum (at this stage it is not even obvious that $C \neq \varnothing$!) we recall that numbers $x \in [0, 1]$ can be represented with a ternary, instead of a decimal, expansion: this means that we can write

$$x = \sum_{i \geq 1} a_i 3^{-i} = 0, a_1 a_2 a_3 \ldots$$

with the ternary digits $a_i \in \{0, 1, 2\}$. As for decimal expansions, this representation is not unique; for instance $1/3$ can be written either as 0.1 or as $0.0222\ldots$, and $2/3$ can be written either as 0.2 or as $0.1222\ldots$. It is easy to check that C_1 corresponds to the set of numbers that can be expressed by a ternary representation not having 1 as first digit, C_2 corresponds to the set of numbers that admit a representation not having 1 as a first or second digit, and so on. It follows that C is the set of numbers that admit a ternary representation not using the digit 1: since the map

$$(a_1, a_2, \ldots) \in \{0, 2\}^{\mathbb{N}^*} \mapsto x = \sum_{i=1}^{\infty} a_i 3^{-i}$$

provides a bijection of $\{0, 2\}^{\mathbb{N}^*}$ with C, and the cardinality of $\{0, 2\}^{\mathbb{N}^*}$ is the continuum, this proves that C has the cardinality of continuum.

Exercise 1.9. Let $\{q_n\}_{n \in \mathbb{N}}$ be an enumeration of the rational numbers in $[0, 1]$, and set

$$A := \bigcup_{n=0}^{\infty} (q_n - \frac{\varepsilon}{4} 2^{-n}, q_n + \frac{\varepsilon}{4} 2^{-n}).$$

Then $A \subset \mathbb{R}$ is open and $\lambda(A) < \sum_n \varepsilon 2^{-n-1} = \varepsilon$ (why is the inequality strict ?). Therefore $[0, 1] \setminus A$ has Lebesgue measure strictly less than ε and an empty interior, because $[0, 1] \setminus A$ does not intersect $\mathbb{Q}$.

Exercise 1.11 Let $\{I_n\}_{n \in \mathbb{N}}$ be an enumeration of the open intervals with rational endpoints of $(0, 1)$. By the construction in Exercise (1.9), for any interval I and any $\delta \in (0, \lambda(I))$ we can find a compact set $C \subset I$ with an empty interior such that $0 < \lambda(C) < \delta$. We will define

$$E := \bigcup_{i=0}^{\infty} C_i$$

where $C_n \subset I_n$ are compact sets with an empty interior, $\lambda(C_n) > 0$ and $\lambda(C_n) < \delta_n$. The choice of C_n and δ_n will be done recursively. Notice first that

$$\lambda(E \cap I_n) \geq \lambda(C_n) > 0 \qquad \forall n \in \mathbb{N},$$

so we have only to take care of the condition $\lambda(E \cap I_n) < \lambda(I_n)$. Set $\beta_n = \lambda(I_n \setminus \cup_0^n C_i)$ and notice that $\beta_n > 0$ because all C_i have an empty interior. Since

$$\lambda(I_n \cap E) \leq \lambda(I_n \cap \bigcup_0^n C_i) + \sum_{i=n+1}^{\infty} \delta_i = \lambda(I_n) - \beta_n + \sum_{i=n+1}^{\infty} \delta_i$$

it suffices to choose δ_n (and C_n) in such a way that $\sum_{n+1}^{\infty} \delta_i < \beta_n$. This is possible, choosing for instance $\delta_{n+1} > 0$ satisfying

$$\delta_{n+1} < \max \left\{ \frac{1}{2}\beta_n, \frac{1}{4}\beta_{n-1}, \ldots, \frac{1}{2^{n+1}}\beta_0 \right\},$$

to get $\delta_i < 2^{n-i}\beta_n$ for $i > n$.

Exercise 1.12. Let A be μ–measurable and let $B, C \in \mathscr{E}$ be satisfying $A \triangle B \subset C$ and $\mu(C) = 0$. For any set $D \subset X$ we have, by monotonicity of μ^*,

$$\mu^*(D \cap A) + \mu^*(D \setminus A) \leq \mu^*(D \cap (B \cup C)) + \mu^*((D \setminus B) \cup C).$$

Since $\mu^*(D \cap C) \leq \mu^*(C) = \mu(C) = 0$, by using twice the subadditivity of μ^* and then the additivity of B we get

$$\mu^*(D \cap A) + \mu^*(D \setminus A) \leq \mu^*(D \cap B) + \mu^*(D \setminus B) = \mu^*(D).$$

Since D is arbitrary, this proves that A is additive.

Exercise 1.13. The statement is trivial if $\mu^*(A) = \infty$. If not, for any $n \in \mathbb{N}^*$ we can find, by the definition of μ^*, a countable union A_n of sets of $\mathscr{N}$ such that $A_n \supset A$ and $\mu(A_n) \leq \mu^*(A) + 1/n$. Then, setting $B := \bigcap_n A_n$ we have $B \supset A$ and $\mu(B) \leq \inf_n \mu^*(A) + 1/n = \mu^*(A)$. The inequality $\mu(B) \geq \mu^*(B)$ follows by the monotonicity of μ^*, taking into account that $\mu^*(B) = \mu(B)$.

Exercise 1.14. $\mathscr{E}_\mu$ is a σ–algebra: stability under complement is immediate, because $A^c \triangle B^c = A \triangle B$; if $A_i \triangle B_i \subset C_i$, then $(\bigcup_i A_i) \triangle (\bigcup_i B_i) \subset \bigcup_i C_i$, and since μ–negligible sets are stable under countable unions, this proves that $\mathscr{E}_\mu$ is stable under countable unions.

The extension $\mu(A) := \mu(B)$, where $B \in \mathscr{E}$ is any set such that $A \triangle B$ is contained in a μ–negligible set of $\mathscr{E}$, is well defined and σ–additive on $\mathscr{E}_\mu$: if $A \triangle B \subset C$ and $A \triangle B' \subset C'$, then $B \triangle B' \subset C \cup C'$; consequently, if $\mu(C) = \mu(C') = 0$ it must be $\mu(B) = \mu(B')$. The σ–additivity can be proven with an argument analogous to the one used to show that $\mathscr{E}_\mu$ is a σ–algebra.

μ–negligible sets of $\mathscr{E}_\mu$ are characterized by the property of being contained in a μ–negligible set of $\mathscr{E}$: if $A \in \mathscr{E}_\mu$ is μ–negligible, there exist μ–negligible sets B, $C \in \mathscr{E}$ with $A \triangle B \subset C$; as a consequence A is contained in the μ–negligible set $B \cup C \in \mathscr{E}$. Conversely, if $A \subset X$ is contained in a μ–negligible set $C \in \mathscr{E}$ we may take $B = \varnothing$ to conclude that $A \in \mathscr{E}_\mu$ and $\mu(A) = 0$.

Exercise 1.15. Let A be additive; by Exercise 1.13 we can find a set $B \in \mathscr{E}$ containing A with $\mu(B) = \mu^*(A)$. The additivity of A and the equality $\mu^*(B) = \mu(B)$ give

$$\mu(B) = \mu^*(A) + \mu^*(B \setminus A).$$

As a consequence $\mu^*(B \setminus A) = 0$. Now we apply Exercise 1.13 again, to find a μ–negligible set $C \in \mathscr{E}$ containing $A \setminus B$. It follows that $A \triangle B$ is contained in C, and therefore A is μ–measurable.

Exercise 1.16. Let us first build a family of pairwise disjoints sets $\{A_i\}_{i \in I} \subset \mathscr{P}(\mathbb{N})$, with I and all sets A_i having an infinite cardinality and $\bigcup_i A_i = \mathbb{N}$ (the construction of the σ–algebra will be more clear if we keep I and $\mathbb{N}$ distinct). The family $\{A_i\}$ can be obtained, for instance, through a bijective correspondence S between $\mathbb{N} \times \mathbb{N}$ and $\mathbb{N}$, setting $A_i := S(\{i\} \times \mathbb{N})$. Then, we define $\pi : \mathbb{N} \to I$ by

$$\pi(n) = i, \text{ where } i \in I \text{ is the unique index such that } n \in A_i$$

and (with the convention $\pi^{-1}(\varnothing) = \varnothing$)

$$\mathscr{F} := \left\{ \pi^{-1}(J) \ : \ J \subset I \right\}.$$

It is immediate to check that $\mathscr{F}$ is a σ–algebra, that $A_i = \pi^{-1}(\{i\}) \in \mathscr{F}$ and that any nonempty set in $\mathscr{F}$ contains one of the sets A_i. Therefore $\mathscr{F}$ contains infinitely many sets, and all of them except $\varnothing$ have an infinite cardinality.

Exercise 1.17. It suffices to define $\mu(A) = 0$ if A has a finite cardinality, and $+\infty$ otherwise. A finite union of sets has an infinite cardinality if and only if at least one of the sets has an infinite cardinality, and this shows that μ is additive.

The solutions of the next exercises require a more advanced knowledge of set theory, and in particular the theory of ordinals, the transfinite induction, the behavior of cardinality under unions and products, and Zorn lemma. We shall denote by ω the smallest uncountable ordinal and by χ the cardinality of continuum.

Exercise 1.18. Notice that $\mathscr{F}^{(j)} \subset \sigma(\mathscr{K})$ implies

$$\left\{ \bigcup_{k=0}^{\infty} A_k, \ B^c \ : \ (A_k) \subset \mathscr{F}^{(j)}, \ B \in \mathscr{F}^{(j)} \right\} \subset \sigma(\mathscr{K}).$$

Therefore, if i is the successor of j, we obtain $\mathscr{F}^{(i)} \subset \sigma(\mathscr{K})$; analogously, if i has no predecessor, and $\mathscr{F}^{(j)} \subset \sigma(\mathscr{K})$ for all $j \in i$, then $\bigcup_{j \in i} \mathscr{F}^{(j)}$, namely $\mathscr{F}^{(i)}$, is contained in $\sigma(\mathscr{K})$. Using these two facts, one obtains by transfinite induction that $\mathscr{F}^{(i)} \subset \sigma(\mathscr{K})$ for all $i \in \omega$. An analogous induction argument shows that $\mathscr{F}^{(i)} \subset \mathscr{F}^{(j)}$ whenever $i \in j$.

So, the union $\mathscr{U} := \bigcup_{i \in \omega} \mathscr{F}^{(i)}$ is contained in $\sigma(\mathscr{K})$ and, to prove that equality holds, it suffices to show that this union is a σ–algebra. Let $(B_k) \subset \mathscr{U}$ and let $i_k \in \omega$ be such that $B_k \in \mathscr{F}^{(i_k)}$. Since i_k are countable and ω is uncountable we have $i := \bigcup_k i_k \in \omega$ and all sets B_k belong to $\mathscr{F}^{(i)}$. It follows that their union belongs to $\mathscr{F}^{(j)}$, where j is the successor of i, and therefore to $\mathscr{U}$. An analogous (and simpler) argument proves that $\mathscr{U}$ is stable under complement.

Exercise 1.19. Obviously $\mathscr{B}(\mathbb{R})$ has at least the cardinality of continuum, so we need only to show an upper bound on the cardinality of $\mathscr{B}(\mathbb{R})$. The proof is based on the fact that a union $\bigcup_{i \in J} X_i$ and a product $\bigtimes_{i \in J} X_i$ have cardinality not greater than χ if the index set J and all sets X_i have cardinality not greater than χ. Let $\mathscr{F}^{(i)}$ be defined as in Exercise 1.19, with $\mathscr{K}$ having at most the cardinality of continuum. Using the previous property of products, with J even countable, one can prove by transfinite induction that, for all $i \in \omega$, $\mathscr{F}^{(i)}$ has at most cardinality χ. If we choose as $\mathscr{K}$ the class of intervals, whose cardinality is (at most) χ, we find

$$\mathscr{B}(\mathbb{R}) = \sigma(\mathscr{K}) = \bigcup_{i \in \omega} \mathscr{F}^{(i)}.$$

Now we use the above mentioned property of unions, with $J = \omega$ and $X_i = \mathscr{F}^{(i)}$, to conclude that $\mathscr{B}(\mathbb{R})$ has at most the cardinality of continuum.

Exercise 1.20. Obviously $\mathscr{L}$ has a cardinality not greater than the cardinality of $\mathscr{P}(\mathbb{R})$; by Bernstein theorem[1] it suffices to show that the cardinality of $\mathscr{P}(\mathbb{R})$ is not greater than the cardinality of $\mathscr{L}$: if C is the Cantor set of Exercise 1.8, we know that $\mathscr{P}(\mathbb{R})$ is in one-to-one correspondence of $\mathscr{P}(C)$, because C has the cardinality of continuum; on the other hand, any subset of C obviously belongs to $\mathscr{L}$, because C has null Lebesgue measure.

Exercise 1.21. Let $\mathscr{E} \subset \mathscr{P}(X)$ be a σ–algebra. Assume by contradiction that $\mathscr{E}$ is infinite and countable. We define the equivalence relation

$$ y \sim y' \qquad \text{if and only if} \quad ((y \in B \iff y' \in B) \quad \forall B \in \mathscr{E}) $$

and let $\mathscr{F}$ be the partition of X in equivalence classes. We now prove that $\mathscr{F} \subset \mathscr{E}$. Indeed, let $F \in \mathscr{F}$, fix $f \in F$, for any $x \notin F$ we have $f \nsim x$ so there must be $B \in \mathscr{E}$ such that $f \in B$ and $x \notin B$ (or the opposite, but then we may consider B^c); given this set B, for any $g \in F$ we have that $g \sim f$ implies $g \in B$, so that $F \subset B$. Since x is arbitrary we conclude that

$$ F = \bigcap_{B \in \mathscr{E}, F \subset B} B. $$

Now, since $\mathscr{E}$ is countable, it follows that $F \in \mathscr{E}$. We eventually note that any set in $\mathscr{E}$ is union of sets in $\mathscr{F}$: but then, if $\mathscr{F}$ were finite then $\mathscr{E}$ would be finite, whereas if $\mathscr{F}$ were infinite then $\mathscr{E}$ would be uncountable.

Exercise 1.22 We define $\mathscr{F}$ as in the solution of the previous exercise, in this case it has finite cardinality, say n; consequently, there are 2^n sets in $\mathscr{E}$.

Exercise 1.23 We define $\mathscr{F}$ as in the solution of Exercise 1.21; we also adapt the above argument to show again that $\mathscr{F} \subset \mathscr{E}$. Indeed, let $F \in \mathscr{F}$, fix $f \in F$, for any $x \notin F$ we have $f \nsim x$ so there must be $B = B_{F,x} \in \mathscr{E}$ such that $f \in B$ and $x \notin B$; and again $F \subset B_{F,x}$. Hence

$$ F = \bigcap_{x \in X, \, x \notin F} B_{F,x} $$

and this proves that $\mathscr{F} \subset \mathscr{A}$, since X is countable. We then use the Axiom of Choice to define a function $\phi : \mathscr{F} \to X$ such that $\phi(F) \in F$, and eventually define $\tilde{\mu} = \sum_{F \in \mathscr{F}} \mu(F) \delta_{\phi(x)}$.

---

[1] If A has cardinality not greater than B, and B has cardinality not greater than A, then there exists a bijection between A and B

Exercise 1.24. We begin our construction with an algebra τ_0 in $\mathscr{P}(\mathbb{N})$ and $\mu_0 : \tau_0 \to \{0, 1\}$ which is additive but not σ–additive. For instance we may take as τ_0 the algebra generated by singletons $\{x\}$ with $x \in \mathbb{N}$ (*i.e.* the sets $A \subset \mathbb{N}$ such that either A or A^c are finite) and set

$$\mu_0(A) := \begin{cases} 0 & \text{if } A \text{ is finite;} \\ 1 & \text{if } A^c \text{ is finite.} \end{cases}$$

We will extend μ_0 to an additive function, that we still denote by μ_0, defined on the whole of $\mathscr{P}(\mathbb{N})$. If such an extension exists, it can't be σ–additive, because $\mu_0(\{n\}) = 0$ for all $n \in \mathbb{N}$, while $\mu_0(\mathbb{N}) = 1$.

In the class $\mathscr{C}$ of pairs (τ, μ) with τ algebra and $\mu : \tau \to \{0, 1\}$ additive, we define the partial order relation $(\tau, \mu) \leq (\tau', \mu')$ by $\tau \subset \tau'$ and $\mu'_{|\tau} = \mu$; then we consider the class $\mathscr{C}_0$ of all (τ, μ) satisfying $(\tau, \mu) \geq (\tau_0, \mu_0)$. By Zorn lemma, we can find a maximal $(\bar{\tau}, \bar{\mu})$ in this class: indeed, it is easy to check that any totally ordered chain $I \subset \mathscr{C}_0$ has an upper bound (τ', μ'), defined by

$$\tau' := \bigcup_{(\tau, \mu) \in I} \tau \quad \text{and} \quad \mu'(A) := \mu(A) \quad \text{where} \quad A \in \tau, (\tau, \mu) \in I.$$

We will show that the maximality of $(\bar{\tau}, \bar{\mu})$ forces $\bar{\tau}$ to coincide with $\mathscr{P}(\mathbb{N})$, so that $\bar{\mu}$ will be the desired extension of μ_0.

Let us assume by contradiction that $\bar{\tau} \subsetneq \mathscr{P}(\mathbb{N})$ and choose $Z \subset \mathbb{N}$ with $Z \notin \bar{\tau}$. We notice that

$$\left\{ (A_1 \cap Z) \cup (A_2 \cap Z^c) \ : \ A_1, A_2 \in \bar{\tau} \right\}$$

is the algebra generated by $\bar{\tau} \cup \{Z\}$. Moreover, either Z or Z^c satisfy the following property

$$\text{for all } A \in \bar{\tau} \text{ with } \bar{\mu}(A) = 1, Z \cap A \neq \varnothing. \tag{A.13}$$

If not, we would be able to find $A_1, A_2 \in \bar{\tau}$ with $A_1 \cap Z = A_2 \cap Z^c = \varnothing$ and $\bar{\mu}(A_1) = \bar{\mu}(A_2) = 1$, so that A_1 and A_2 would be disjoint and $\bar{\mu}(A_1 \cup A_2) = 2$, contradicting the fact that $\bar{\mu}$ maps $\bar{\tau}$ into $\{0, 1\}$. Possibly replacing Z by its complement we shall assume that Z fulfils (A.13).

Now we extend $\bar{\mu}$ to the algebra generated by $\bar{\tau} \cup \{Z\}$, as follows:

$$\tilde{\mu}(B) := \bar{\mu}(A_1) \text{ whenever } A_1, A_2 \in \bar{\tau} \text{ and } B = (A_1 \cap Z) \cup (A_2 \cap Z^c). \tag{A.14}$$

Let us check that $\tilde{\mu}$ is well defined and additive.

1. $\tilde{\mu}$ is well defined: if

$$B = (A_1 \cap Z) \cup (A_2 \cap Z^c) = (A_3 \cap Z) \cup (A_4 \cap Z^c)$$

   then $(A_1 \cap Z) = (A_3 \cap Z)$, and if $\bar{\mu}(A_1) \neq \bar{\mu}(A_3)$ then one of the two numbers, say $\bar{\mu}(A_1)$, equals 1, while $\bar{\mu}(A_3) = 0$. Defining $A := A_1 \setminus A_3$ we have $\bar{\mu}(A) = 1$ and $A \cap Z = \varnothing$, contradicting (A.13).
2. Suppose B, $B' \in \bar{\tau}$ are disjoint. Let $B = (A_1 \cap Z) \cup (A_2 \cap Z^c)$ and $B' = (A_1' \cap Z) \cup (A_2' \cap Z^c)$. Then $A_1 \cap A_1' \cap Z = \varnothing$. Setting $A_1'' := A_1' \setminus A_1$ we still have $B' = (A_1'' \cap Z) \cup (A_2' \cap Z^c)$, and then we can use the additivity of $\bar{\mu}$ to conclude that

$$\tilde{\mu}(B \cup B') = \bar{\mu}(A_1 \cup A_1'') = \bar{\mu}(A_1) + \bar{\mu}(A_1'') = \tilde{\mu}(B) + \tilde{\mu}(B').$$

If $B \in \tau$ we can choose $A_1 = A_2 = B$ in (A.14) to obtain that $\tilde{\mu}(B) = \bar{\mu}(B)$, so that $\tilde{\mu}$ extends $\bar{\mu}$ to the algebra generated by $\bar{\tau} \cup \{Z\}$. This violates the maximality of $(\bar{\tau}, \bar{\mu})$.

Exercise 1.25 We obviously need only to show that the cardinality of C is at least equal to the continuum. By the inner regularity of λ we can assume with no loss of generality that C is closed. Now, we define $A = (0, 1) \setminus C$ and

$$g(t) := \lambda\big([0, t] \cap C\big) \qquad t \in [0, 1].$$

This continuous function maps continuously $[0, 1]$ onto $[0, \lambda(C)]$, and it is constant in any connected component of A, so that $g(A)$ is at most countable. Since $g(C)$ contains $[0, \lambda(C)] \setminus g(A)$ we obtain that C has cardinality at least equal to the continuum (one can actually see that $g(C) = g([0, 1])$).

Exercise 1.26 Since K is totally bounded, for all $\epsilon > 0$ there exist finitely many balls $B_1, \ldots, B_N$ with radius ϵ whose union covers K. The properties of μ imply the existence of an index i such that $\mu(\{n : x_n \in B_i\}) = 1$. Now we start with $\epsilon = 1$ and find a closed ball $B^{(1)}$ with radius 1 such that $\mu(\{n : x_n \in B^{(1)}\}) = 1$. Repeating this construction in $B^{(1)}$ we find a closed ball $B^{(2)}$ with radius $1/2$ contained in $B^{(1)}$ with $\mu(\{n : x_n \in B^{(2)}\}) = 1$. Continuing in this way, if z is the common point of the balls $B^{(i)}$, we find x_n μ-converges to z.

Chapter 2

Exercise 2.1 The verification is straightforward and is omitted.

Exercise 2.2 Let φ, $\psi : X \to \mathbb{R}$ be $\mathscr{E}$–measurable. If $\varphi(x) + \psi(x) < t$ we can find a rational number r such that $\varphi(x) < r$ and $\psi(x) < t - r$, hence

$$\{\varphi + \psi < t\} = \bigcup_{r \in \mathbb{Q}} [\{\varphi < r\} \cap \{\psi < t - r\}].$$

This proves that $\varphi + \psi$ is $\mathscr{E}$–measurable. Analogously, since

$$\{\varphi^2 > a\} = \{\varphi > \sqrt{a}\} \cup \{\varphi < -\sqrt{a}\}, \quad a \geq 0$$

we obtain that φ^2 is measurable. Considering the difference $(\varphi + \psi)^2 - (\varphi - \psi)^2$ we obtain that $\varphi\psi$ is $\mathscr{E}$–measurable.

Exercise 2.3. (i) The verification of the axioms of distance is immediate. In order to prove the compactness of $\overline{\mathbb{R}}$, let us consider a sequence $(x_n) \subset \overline{\mathbb{R}}$. If $\sup_n x_n = +\infty$ we can find for any k an index $n(k)$ such that $x_{n(k)} \geq k$; it follows that $d(x_{n_k}, +\infty) = |\arctan x_{n(k)} - \pi/2|$ tends to 0, so that $x_{n(k)} \to +\infty$ in the metric space. Analogously, if $\inf_n x_n = -\infty$ we can find a subsequence converging to $-\infty$ in $(\overline{\mathbb{R}}, d)$. Finally, if both $\sup_n x_n$ and $\inf_n x_n$ are finite, the sequence (x_n) is bounded and we can extract, thanks to the Bolzano–Weierstrass theorem, a subsequence $x_{n(k)}$ converging to $x \in \mathbb{R}$. The continuity of $z \mapsto \arctan z$ implies that $x_{n(k)} \to x$ in $(\overline{\mathbb{R}}, d)$. To prove the equivalence of the two topologies, let us work with closed sets: if $C \subset \mathbb{R}$ is closed with respect to the $(\overline{\mathbb{R}}, d)$ topology, then it is closed with respect to the Euclidean topology, because $|x_n - x| \to 0$ implies $|\arctan x_n - \arctan x| \to 0$. On the other hand, if $|\arctan x_n - \arctan x| \to 0$ then for n large enough $\arctan x_n$ belongs to an interval $I := (\arctan x - \varepsilon, \arctan x + \varepsilon) \subset (-\pi/2, \pi/2)$; the continuity of $y \mapsto \tan y$ in I implies that $x_n \to x$. This proves the converse implication, and the equivalence of the two topologies.

(ii) We notice first that, according to (i), $\mathscr{B}(\mathbb{R})$ and $\{-\infty\}$, $\{+\infty\}$ belong to $\mathscr{B}(\overline{\mathbb{R}})$. Therefore, if f is measurable between $\mathscr{E}$ and the Borel σ–algebra of $(\overline{\mathbb{R}}, d)$, then it is $\mathscr{E}$–measurable according to (2.2). According to the measurability criterion, in order to prove the converse implication it suffices to show that $\mathscr{B}(\overline{\mathbb{R}})$ is generated by $\mathscr{B}(\mathbb{R}) \cup \{-\infty\} \cup \{+\infty\}$: this follows by the fact that if $C \subset \overline{\mathbb{R}}$ is closed, then

$$C = (C \cap \mathbb{R}) \cup (C \cap \{-\infty\}) \cup (C \cap \{+\infty\})$$

(again by (i)) belongs to the algebra generated by $\mathscr{B}(\mathbb{R}) \cup \{-\infty\} \cup \{+\infty\}$, therefore the σ–algebra generated by this family of sets contains $\mathscr{B}(\overline{\mathbb{R}})$.

Exercise 2.4. If $\{f \neq g\}$ is contained in a μ–negligible set C of $\mathscr{E}$, for some $\mathscr{E}$–measurable function g, then $\{f > t\} \Delta \{g > t\} \subset C$ for all $t \in \mathbb{R}$, and since $\{g > t\} \in \mathscr{E}$ it follows that $\{f > t\} \in \mathscr{E}_\mu$; this means that f is $\mathscr{E}_\mu$–measurable. Conversely, assume that f is $\mathscr{E}_\mu$–measurable and find for all $q \in \mathbb{Q}$ a set $B_q \in \mathscr{E}$ and a μ–negligible set $C_q \in \mathscr{E}$ with $\{f > q\} \Delta B_q \subset C_q$. We define

$$g(x) := \sup \{q \in \mathbb{Q} : x \in B_q\}, \qquad C := \bigcup_{q \in \mathbb{Q}} C_q.$$

Since $\{g \leq t\} = \bigcap_{q \leq t} B_q$ we have that g is $\mathscr{E}$–measurable. Let us prove that $f(x) = g(x)$ for all $x \notin C$: for any such x we have $x \in B_q$ for all $q < f(x)$, therefore $g(x) \geq f(x)$; if the inequality were strict, there would exist $q \in \mathbb{Q}$ with $x \in B_q$ and $q > f(x)$, therefore x would be in $B_q \setminus \{f > q\} \subset C_q \subset C$.

Exercise 2.5. If $\sigma \leq \tau$ we can find a nondecreasing family of partitions $\sigma_1, \ldots, \sigma_n$ with $\sigma_1 = \sigma$, $\sigma_n = \tau$ and $\sigma_{i+1} \setminus \sigma_i$ containing just one point. Therefore, in the proof of the monotonicity of $\sigma \mapsto I_\sigma(f)$ we need only to show that $I_\sigma(f) \leq I_{\sigma \cup \{t\}}(f)$ whenever $t \in (0, \infty) \setminus \sigma$. Let $\sigma = \{t_0, \ldots, t_N\}$ and let i be the last index such that $t_i < t$. If $i < N$ we use the inequality

$$\begin{aligned}(t_{i+1} - t_i) f(t_{i+1}) &= (t_{i+1} - t) f(t_{i+1}) + (t - t_i) f(t_{i+1}) \\ &\leq (t_{i+1} - t) f(t_{i+1}) + (t - t_i) f(t)\end{aligned}$$

adding to both sides $\sum_{j \neq i} (t_{j+1} - t_j) f(t_{j+1})$ we obtain $I_\sigma(f) \leq I_{\sigma \cup \{t\}}(f)$. If $i = N$ the argument is even easier, because the difference $I_{\sigma \cup \{t\}}(f) - I_\sigma(f)$ is given by $(t - t_N) f(t)$.

Now, let $f, g : (0, +\infty) \to [0, +\infty)$ be given; since $I_\sigma(f + g) = I_\sigma(f) + I_\sigma(g)$ we get $I_\sigma(f + g) \leq \int_0^\infty f(t)\, dt + \int_0^\infty g(t)\, dt$. Since $\sigma \in \Sigma$ is arbitrary, this proves that

$$\int_0^\infty f(t) + g(t)\, dt \leq \int_0^\infty f(t)\, dt + \int_0^\infty g(t)\, dt.$$

In order to prove the converse inequality, fix $L < \int_0^\infty f(t)\, dt$, $M < \int_0^\infty g(t)\, dt$ and find $\sigma, \eta \in \Sigma$ with $I_\sigma(f) > L$ and $I_\eta(g) > M$; then

$$\int_0^\infty f(t) + g(t)\, dt \geq I_{\sigma \cup \eta}(f + g) = I_{\sigma \cup \eta}(f) + I_{\sigma \cup \eta}(g)$$

$$\geq I_\sigma(f) + I_\eta(g) > L + M.$$

Letting $L \uparrow \int_0^\infty f(t)\, dt$ and $M \uparrow \int_0^\infty g(t)\, dt$ the inequality is proved.

Exercise 2.6. We will prove that f_* is lower semicontinuous, the proof of the upper semicontinuity of f^* being analogous. Let $(x_n) \subset \mathbb{R}$ be converging to x and use the definition of $f_*(x_n)$ to find $y_n \in \mathbb{R}$ such that

$$|x_n - y_n| < \frac{1}{n} \qquad \text{and} \qquad f(y_n) \leq f_*(x_n) + \frac{1}{n}.$$

Then (y_n) still converges to x, so that

$$f_*(x) \leq \liminf_{n \to \infty} f(y_n) \leq \liminf_{n \to \infty} f_*(x_n) + \frac{1}{n} = \liminf_{n \to \infty} f_*(x_n).$$

Exercise 2.7. Let $t \in \mathbb{R}$ and let $(x_n) \subset \{f_* \leq t\}$ be convergent to x. Then, the lower semicontinuity of f_* gives

$$f_*(x) \leq \liminf_{n \to \infty} f_*(x_n) \leq t.$$

This proves that $x \in \{f_* \leq t\}$, so that $\{f_* \leq t\}$ is closed. The proof for f^* is similar. Since the Borel σ–algebra is generated by halflines, it follows that f^* and f_* are Borel, and the same is true for the set $\{f_* = f^*\}$, that coincides with Σ.

Exercise 2.8. Set $\varphi_0 := \varphi$, $A_0 := \{\varphi_0 \geq a_0\}$ and $\varphi_1 := \varphi - a_0 \mathbb{1}_{A_0} \geq 0$. Then, set $A_1 := \{\varphi_1 \geq a_1\}$ and $\varphi_2 := \varphi_1 - a_1 \mathbb{1}_{A_1}$ and so on. If $\varphi(x) = +\infty$ then $\varphi_n(x) = +\infty$ for all n, so that x belongs to all sets A_i and $\sum_{i=0}^n a_i \mathbb{1}_{A_i}(x) = +\infty$. We then assume that $\varphi(x) < +\infty$ in the following. By construction we have that $0 \leq \varphi_{i+1} \leq \varphi_i \leq \cdots \leq \varphi_0 = \varphi$, hence

$$\varphi = \varphi_{n+1} + \sum_{i=0}^n (\varphi_i - \varphi_{i+1}) = \varphi_{n+1} + \sum_{i=0}^n a_i \mathbb{1}_{A_i}.$$

This proves that $\varphi \geq \sum_i a_i \mathbb{1}_{A_i}$. If the inequality were strict for some $x \in X$ with $\varphi(x) < +\infty$, we could find $\varepsilon > 0$ such that $\varphi_i(x) \geq \varepsilon$ for all $i \in \mathbb{N}$, and since $a_i < \varepsilon$ for i large enough, we would get $x \in A_i$ for i large enough. But since the series $\sum_i a_i$ is not convergent, we would get $\sum_i a_i \mathbb{1}_{A_i}(x) = \infty$, a contradiction.

Exercise 2.9. Assume by contradiction that the absolute continuity property fails. Then, for some $\varepsilon > 0$ we can find A_i with $\mu(A_i) < 2^{-i}$ and $\int_{A_i} |\varphi| \, d\mu \geq \varepsilon$. It follows that the set $B := \limsup_i A_i$ is μ–negligible, and

$$B_n := \bigcup_{i \geq n} A_i \setminus B \downarrow \varnothing.$$

Since $\int_{B_n} |\varphi| \, d\mu \geq \int_{A_n} |\varphi| \, d\mu \geq \varepsilon$ we find a contradiction with the dominated convergence theorem applied to the functions $\mathbb{1}_{B_n} |\varphi|$, pointwise converging to 0.

Exercise 2.10. Let $\varepsilon > 0$ be given and let $\delta > 0$ be such that $\int_A |\varphi| \, d\mu < \varepsilon/2$ whenever $A \in \mathcal{E}$ and $\mu(A) < \delta$. The triangle inequality gives, with the same choice of A, $\int_A |\varphi_n| \, d\mu < \varepsilon$ for $n > n_0$, provided $\|\varphi_n - \varphi\|_1 < \varepsilon/2$ for $n > n_0$. Since $\varphi_1, \ldots, \varphi_{n_0}$ are integrable, we can find $\delta_i > 0$ such that $\int_A |\varphi_i| \, d\mu < \varepsilon$ whenever $A \in \mathcal{E}$ and $\mu(A) < \delta_i$. If

$\delta_0 = \min\{\delta, \min_i \delta_i\}$, we have $\int_A |\varphi_n|\, d\mu < \varepsilon/2$ whenever $n \in \mathbb{N}$, $A \in \mathscr{E}$ and $\mu(A) < \delta$.

A possible example for the second question is $\Omega = [0, 1]$, $\mu = \lambda$ the Lebesgue measure, and $\varphi_n = \frac{2^n}{n}\mathbb{1}_{[2^{-n}, 2^{1-n})}$. The uniform integrability is a direct consequence of the convergence of φ_n to 0 in L^1. If $\varphi_n \leq g$, then

$$\sum_{n=1}^{\infty} \varphi_n = \sum_{n=1}^{\infty} \varphi_n \leq g$$

but $\int \sum_{n=1}^{\infty} \varphi_n = \sum_1^{\infty} 1/n = +\infty$.

Exercise 2.11. (a) For any $y \in X$ we have

$$g_\lambda(x) \leq g(y) + \lambda d(x, y) \leq g(y) + \lambda d(x', y) + \lambda d(x, x').$$

Since y is arbitrary we get $g_\lambda(x) \leq g_\lambda(x') + \lambda d(x, x')$. Reversing the roles of x and x' the inequality is achieved.

(b) Clearly the family (g_λ) is monotone with respect to λ, and since we can always choose $y = x$ in the minimization problem we have $g_\lambda(x) \leq g(x)$. Assume that $\sup_\lambda g_\lambda(x)$ is finite (otherwise the statement is trivial) and let x_λ such that $g_\lambda(x) + \lambda^{-1} \geq g(x_\lambda) + \lambda d(x, x_\lambda)$. This inequality implies that $x_\lambda \to x$ as $\lambda \to \infty$ and, now neglecting the term $\lambda d(x, x_\lambda)$, that

$$g_\lambda(x) + \frac{1}{\lambda} \geq g(x_\lambda).$$

Passing to the limit in this inequality as $\lambda \to \infty$ and using the lower semicontinuity of g we get $\sup_\lambda g_\lambda(x) \geq g(x)$.

Exercise 2.12. Let us first assume that f is bounded. For $\varepsilon > 0$ we consider the functions

$$f_\varepsilon(x, y) := \frac{1}{2\varepsilon} \int_{x-\varepsilon}^{x+\varepsilon} f(x', y)\, dx'.$$

Since $x \mapsto f(x, y)$ is continuous, we can apply the mean value theorem to obtain that $f_\varepsilon(x, y) \to f(x, y)$ as $\varepsilon \downarrow 0$. So, in order to show that f is a Borel function, we need only to show that f_ε are Borel.

We will prove indeed that f_ε are continuous: let $x_n \to x$ and $y_n \to y$; since $f(x', y_n) \to f(x', y)$ for all $x' \in \mathbb{R}$, we have

$$\mathbb{1}_{[x_n-\varepsilon, x_n+\varepsilon]}(x') f(x', y_n) \to \mathbb{1}_{[x-\varepsilon, x+\varepsilon]}(x') f(x', y)$$

for all $x' \in \mathbb{R} \setminus \{x - \varepsilon, x + \varepsilon\}$. Therefore, since f is bounded, the dominated convergence theorem yields

$$f_\varepsilon(x, y) = \frac{1}{2\varepsilon} \int_{\mathbb{R}} \mathbb{1}_{[x-\varepsilon, x+\varepsilon]}(x') f(x', y)\, dx'$$

$$= \frac{1}{2\varepsilon} \lim_{n\to\infty} \int_{\mathbb{R}} \mathbb{1}_{[x_n-\varepsilon, x_n+\varepsilon]}(x') f(x', y_n)\, dx' = \lim_{n\to\infty} f_\varepsilon(x_n, y_n).$$

In the general case when f is not bounded we approximate it by the bounded functions $f_h(x) := \max\{-h, \min\{f(x), h\}\}$, with $h \in \mathbb{N}$, that are still separately continuous, and therefore Borel.

Chapter 3

Exercise 3.1. On the real line, endowed with the Lebesgue measure, the function $(1 + |x|)^{-1}$ belongs to L^2, but not to L^1, and the function $|x|^{-1/2}\mathbb{1}_{(0,1)}(x)$ belongs to L^1, but not to L^2. Turning back to the general case, if $\varphi \in L^{p_1} \cap L^{p_2}$ with $p_1 \leq p_2$, from the inequality

$$|\varphi|^p \leq \max\{|\varphi|^{p_1}, |\varphi|^{p_2}\} \leq |\varphi|^{p_1} + |\varphi|^{p_2} \qquad \forall p \in [p_1, p_2]$$

(that can be verified considering separately the cases $|\varphi| \leq 1$ and $|\varphi| > 1$) we get that $\varphi \in L^p$ for all $p \in [p_1, p_2]$.

Exercise 3.2. The statement is trivial if $\|f\|_q = 0$, so we assume that $\|f\|_q > 0$. For $\epsilon > 0$ the set $X_\epsilon := \{|f| > \epsilon\}$ has finite μ–measure, by the Markov inequality, hence the inclusion between L^r spaces for finite measures gives that $|f|\mathbb{1}_{X_\epsilon} \in L^p(X, \mathcal{E}, \mu)$. Since the dominated convergence theorem gives

$$\lim_{\epsilon \downarrow 0} \int_X |f - f\mathbb{1}_{X_\epsilon}|^q \, d\mu = \lim_{\epsilon \downarrow 0} \int_{X \setminus X_\epsilon} |f|^q \, d\mu = 0$$

we can choose $\tilde{f} = f\mathbb{1}_{X_\epsilon}$ for $\epsilon > 0$ small enough.

Exercise 3.3. By homogeneity we can assume that $\|\varphi\|_p = 1$ and $\|\psi\|_q = 1$. Since

$$\int_X \left(\frac{|\varphi|^p}{p} + \frac{|\psi|^q}{q} - |\varphi||\psi| \right) d\mu = \frac{\|\varphi\|_p}{p} + \frac{\|\psi\|_q}{q} - 1 = 0$$

and the function among parentheses is nonnegative, it follows that if vanishes μ–a.e. In particular, for μ–a.e. x, $|\varphi(x)|$ is a minimizer of

$$y \mapsto \frac{y^q}{q} - |\psi(x)|y$$

in $[0, +\infty)$. But this problem has a unique minimizer, given by $|\psi(x)|^{q-1}$, and we conclude.

Exercise 3.4. It suffices to apply Hölder's inequality to the functions $|\varphi|^r$ and $|\psi|^r$, with the dual exponents p/r and q/r, to obtain

$$\|\varphi\psi\|_r^r \leq \||\varphi|^r\|_{p/r} \||\psi|^r\|_{q/r} = \|\varphi\|_p^r \|\psi\|_q^r.$$

Exercise 3.5. The positive part and the negative part of $\varphi - \varphi_n$ have the same integral, hence

$$\int_X |\varphi - \varphi_n|\, d\mu = 2 \int_X (\varphi - \varphi_n)^+\, d\mu.$$

The condition $\liminf_n \varphi_n \geq \varphi$ ensures that $(\varphi - \varphi_n)^+$ is pointwise convergent to 0; in addition, since φ_n are nonnegative, the functions are dominated by φ^+. Therefore the dominated convergence theorem gives the result.

Exercise 3.6. If $\psi_n \to \psi$ μ–a.e. we apply Fatou's lemma to the functions $\psi_n + \varphi_n$ to obtain

$$\liminf_{n \to \infty} \int_X \psi_n + \varphi_n\, d\mu \geq \int_X \psi + \varphi\, d\mu.$$

Therefore

$$\limsup_{n \to \infty} \int_X \psi_n\, d\mu + \liminf_{n \to \infty} \int_X \varphi_n\, d\mu \geq \int_X \varphi\, d\mu + \int_X \psi\, d\mu.$$

Subtracting $\int \psi\, d\mu$ from both sides the statement is achieved. In the general case, let $n(k)$ be a subsequence such that $\lim_k \int \varphi_{n(k)}\, d\mu = \liminf_n \int_X \varphi_n$, and let $n(k(s))$ be a further subsequence converging to φ μ–a.e. Then

$$\liminf_{n \to \infty} \int_X \varphi_n\, d\mu = \lim_{s \to \infty} \int_X \varphi_{n(k(s))}\, d\mu \geq \int_X \liminf_n \varphi_{n(k(s))}\, d\mu$$

$$\geq \int_X \liminf_{n \to \infty} \varphi_n\, d\mu.$$

Exercise 3.7. We show only how (3.13) implies $g(tx + (1-t)y) \leq tg(x) + (1-t)g(y)$ for all $x,\ y \in J$ and $t \in [0,1]$. We prove first, by induction on m, that

$$g\left(\sum_{i=1}^{2^m} \frac{1}{2^m} x_i\right) \leq \sum_{i=1}^{2^m} \frac{1}{2^m} g(x_i)$$

for all $x_1, \ldots, x_{2^m} \in J$. The case $m = 1$ is (3.13) and the induction step can be achieved grouping the terms as follows:

$$\sum_{i=1}^{2^m} \frac{1}{2^m} x_i = \frac{1}{2}\left(\sum_{i=1}^{2^{m-1}} \frac{1}{2^{m-1}} x_i + \sum_{i=1}^{2^{m-1}} \frac{1}{2^{m-1}} x_{2^{m-1}+i}\right).$$

Now, considering the case when $x_i = x$ for $1 \leq i \leq k$ and $x_i = y$ otherwise, we get

$$g(tx + (1 - t)y) \leq tg(x) + (1 - t)g(y) \qquad \text{with } t = \frac{k}{2^m}.$$

Since g is continuous, by approximation we get $g(tx + (1 - t)y) \leq tg(x) + (1 - t)g(y)$ for all x, $y \in J$ and $t \in [0, 1]$.

Exercise 3.8. Let us first show the existence of z_0. Let $A = g(\mathbb{R})$ and let $u_n = g(z_n)$ with $u_n \downarrow \inf A$. Since u_n is uniformly bounded from above, our assumption on g ensures that (z_n) is bounded. By the Bolzano-Weierstrass theorem we can find a subsequence $z_{n(k)}$ convergent to $z \in \mathbb{R}$. The continuity of g gives that $u_{n(k)} = g(z_{n(k)})$ converge to $g(z)$. It follows that $\inf A$ is finite and coincides with $g(z)$. Now, by applying the convexity inequality of the previous exercise with $x = z_2$, $y = z_0$ and $t = (z_1 - z_0)/(z_2 - z_0)$, we get

$$\frac{g(z_2) - g(z_1)}{z_2 - z_1} \geq \frac{g(z_1) - g(z_0)}{z_1 - z_0} \geq 0$$

for $z_0 < z_1 < z_2$, proving the monotonicity of g in $[z_0, +\infty)$. The argument in $(-\infty, z_0]$ is analogous.

Exercise 3.9. Fatou's lemma gives $\liminf_n \int \varphi_n \, d\mu \geq \int \liminf_n \varphi_n \, d\mu \geq \int \varphi \, d\mu$. Therefore $t_n := \int \varphi_n \, d\mu \to t := \int \varphi \, d\mu$; we can apply Exercise 3.5 to the functions φ_n / t_n to obtain that $\varphi_n / t_n \to \varphi / t$ in L^1. From this, taking into account that $t_n \to t$, the convergence of φ_n to φ in L^1 follows.

Exercise 3.10. Let $\Psi(c) := \Phi(c)/c$ and notice that $|\varphi_i| \leq c\Phi(|\varphi_i|)/\Phi(c) = \Phi(|\varphi_i|)/\Psi(c)$ on $\{|\varphi_i| \geq c\}$. Therefore

$$\int_A |\varphi_i| \, d\mu \leq \int_{A \cap \{|\varphi_i| \geq c\}} \frac{\Phi(|\varphi_i|)}{\Psi(c)} \, d\mu + \int_{A \cap \{|\varphi_i| < c\}} |\varphi_i| \, d\mu \leq \frac{M}{\Psi(c)} + c\mu(A).$$

Let us choose c sufficiently large, such that $M/\Psi(c) < \varepsilon/2$, and then $\delta > 0$ such that $c\delta < \varepsilon/2$. The inequality above yields $\int_A |\varphi_i| \, d\mu < \varepsilon$ whenever $\mu(A) < \delta$.

Exercise 3.11. Let $(f_n) \subset C_b(X)$ be converging in L^1 to f, and let $f_{n(k)}$ be a subsequence pointwise convergent μ–a.e. to f. Then, given any $\varepsilon > 0$, by Egorov theorem we can find a Borel set $B \subset X$ with $\mu(B) < \varepsilon$ and $f_{n(k)} \to f$ uniformly on B^c. By the inner regularity of the measure we can find a closed set $C \subset B^c$ such that $\mu(X \setminus C) < \varepsilon$. The function f restricted to C, being the uniform limit of bounded continuous functions, is bounded and continuous.

Chapter 4

Exercise 4.1. Notice that $\langle \cdot, \cdot \rangle$ is obviously symmetric, that $\langle x, -y \rangle = -\langle x, y \rangle = \langle -x, y \rangle$ and that $\langle x, x \rangle = \|x\|^2 \geq 0$, with equality only if $x = 0$. Notice that the parallelogram identity gives

$$\|x + x' + 2y\|^2 + \|x - x'\|^2 = 2\|x + y\|^2 + 2\|x' + y\|^2$$
$$= 8\langle x, y \rangle + 8\langle x', y \rangle - 2\|x - y\|^2 - 2\|x' - y\|^2$$

and

$$\|x + x' - 2y\|^2 + \|x - x'\|^2 = 2\|x - y\|^2 + 2\|x' - y\|^2$$
$$= 8\langle x, -y \rangle + 8\langle x', -y \rangle - 2\|x + y\|^2 - 2\|x' + y\|^2.$$

Subtracting and dividing by 4 we get

$$\langle x + x', 2y \rangle = 4\langle x, y \rangle + 4\langle x', y \rangle - 2\langle x, y \rangle - 2\langle x', y \rangle.$$

So, we proved that $\langle x + x', 2y \rangle = 2\langle x, y \rangle + 2\langle x', y \rangle$. Using the relation $\langle u, 2v \rangle = 4\langle u/2, v \rangle$ (due to the definition of $\langle \cdot, \cdot \rangle$ and the homogeneity of $\| \cdot \|$), we get

$$\left\langle \frac{x + x'}{2}, y \right\rangle = \frac{1}{2}\langle x, y \rangle + \frac{1}{2}\langle x', y \rangle.$$

Setting $x = t_1 v$, $x' = t_2 v$, and defining the continuous function $\phi(t) = \langle tv, y \rangle$, we get

$$\phi\left(\frac{t_1 + t_2}{2}\right) = \frac{1}{2}\phi(t_1) + \frac{1}{2}\phi(t_2).$$

This means that ϕ and $-\phi$ are convex in $\mathbb{R}$, so that ϕ is an affine function, and since $\phi(0) = 0$ we get $\phi(t) = t\phi(0)$, i.e. $\langle tu, y \rangle = t\langle u, y \rangle$. Coming back to the identity above, we get $\langle x + x', y \rangle = \langle x, y \rangle + \langle x', y \rangle$.

Exercise 4.2. Assume that $y = \pi_K(x)$. For all $z \in K$ and $t \in [0, 1]$ we have $y + t(z - y)$ belongs to K, so that

$$\|y + t(z - y) - x\|^2 \geq \|y - z\|^2.$$

Expanding the squares we get

$$t^2\|z - y\|^2 + 2t\langle z - y, y - x \rangle \geq 0 \qquad \forall t \in [0, 1].$$

This implies (either dividing by $t > 0$ and passing to the limit as $t \downarrow 0$, or computing the right derivative at $t = 0$) that $\langle z - y, x - y \rangle \leq 0$. Conversely, if for some $y \in K$ this condition holds for all $z \in K$, the argument can be reversed to get $\|y + t(z - y) - x\| \geq \|y - x\|$ for all $t \geq 0$. Choosing $t = 1$ we get $\|z - x\| \geq \|y - x\|$, proving that $y = \pi_K(x)$.

Exercise 4.3. Let Y_k be the vector space spanned by $\{f_1, \ldots, f_k\}$ and let us prove by induction on $k \geq 1$ that f_i is orthogonal to f_j whenever $1 \leq i < j \leq k$. First we observe that if this property holds for some k, then Y_k is k-dimensional and coincides with the vector space spanned by $\{v_1, \ldots, v_k\}$ (being contained in it, and with the same dimension).

The orthogonality of the vectors f_i can be obtained just noticing that

$$f_k = v_k - \sum_{i=1}^{k-1} \langle v_k, e_i \rangle e_i.$$

So, $f_k = v_k - \pi_{Y_{k-1}}(v_k)$ is orthogonal to all vectors in Y_{k-1}. It follows that $\langle e_k, e_i \rangle = 0$ for all $i < k$.

Exercise 4.4. Let $y = x - \sum_k \langle x, e_k \rangle e_k$; we know that the series converges in H by Bessel's inequality. In order to show that $\sum_k \langle x, e_k \rangle e_k = \pi_X(x)$ it suffices to prove that y is orthogonal to all vectors in X. But since any vector $v \in X$ can be represented as a series, it suffices to show that $\langle v, e_i \rangle = 0$ for all i. The continuity and linearity of the scalar product give

$$\langle y, e_i \rangle = \langle x, e_i \rangle - \sum_{k=0}^{\infty} \langle x, e_k \rangle \langle x, e_i \rangle = \langle x, e_i \rangle - \langle x, e_i \rangle = 0.$$

Exercise 4.5 Since X and its scalar product coincide with $L^2([0, 1], \mathscr{P}([0, 1]), \mu)$, where μ is the counting measure in $[0, 1]$, we obtain that X is an Hilbert space. Let us prove by contradiction that X is not separable. If $S = \{f_n\}_{n \geq 1}$ were a dense subset, it could be possible to find a countable set $D \subset [0, 1]$ such that $f_n(x) = 0$ for all n and all $x \in [0, 1] \setminus D$. Since $[0, 1]$ is not countable we can find $x_0 \in [0, 1] \setminus D$ and define $g_0(x)$ equal to 1 if $x = x_0$ and equal to 0 if $x \neq x_0$. We claim that g_0 does not belong to the closure of S. If this property fails, we can find a sequence $(f_{n(k)}) \subset S$ convergent to g_0 μ–a.e. in $[0, 1]$; but, convergence μ–a.e. corresponds to pointwise convergence and since $g_0(x_0) \neq 0$, while $f_{n(k)}(x_0) = 0$ for all k, we obtain a contradiction.

Exercise 4.6. By Parseval identity we know that $x \mapsto (\langle x, e_i \rangle)$ is a linear isometry from H to ℓ_2. As a consequence, taking the parallelogram identity into account, the scalar product is preserved.

Exercise 4.7. We consider the class of orthonormal systems $\{e_i\}_{i \in I}$ of H, ordered by inclusion. Zorn's lemma ensures the existence of a maximal system $\{e_i\}_{i\in I}$. Let V be the subspace spanned by e_i, let Y be its closure (still a subspace) and let us prove that $Y = H$. Indeed, if Y were a proper subspace of H, we would be able to find, thanks to Corollary 4.5, a unit vector e orthogonal to all vectors in Y, and in particular to all vectors e_i. Adding e to the family $\{e_i\}_{i\in I}$ the maximality of the family would be violated. Now, by the just proved density of V in H, given any $x \in H$ we can find a sequence of vectors (v_n), finite combinations of vectors e_i, such that $\|x - v_n\| \to 0$. If we denote by $J_n \subset I$ the set of indexes used to build the vectors $\{v_1, \ldots, v_n\}$, and by H_n the vector space spanned by $\{e_i\}_{i\in J_n}$, we know by Proposition 4.6 that

$$\|x - \sum_{i\in J_n}\langle x, e_i\rangle e_i\| \leq \|x - v_n\| \to 0.$$

As a consequence, setting $J = \cup_n J_n$, we have $x = \sum_{i\in J}\langle x, e_i\rangle e_i$.

Chapter 5

Exercise 5.1. The functions $\sin mx \cos lx$ are odd, therefore their integral on $(-\pi, \pi)$ vanishes. To show that $\sin mx$ is orthogonal to $\sin lx$ when $l \neq m$, we integrate twice by parts to get

$$\int_{-\pi}^{\pi} \sin mx \sin lx \, dx = \frac{m}{l} \int_{-\pi}^{\pi} \cos mx \cos lx \, dx$$
$$= \frac{m^2}{l^2} \int_{-\pi}^{\pi} \sin mx \sin lx \, dx.$$

The integrals of products $\cos mx \cos lx$ can be handled analogously.

Exercise 5.2. Since for $N < M$ we have

$$\|\sum_{n=0}^{N} x_n - \sum_{n=0}^{M} x_n\| \leq \sum_{i=N+1}^{M} \|x_i\| \leq \sum_{i=N+1}^{\infty} \|x_i\|$$

we obtain that $(\sum_0^N x_i)$ is a Cauchy sequence in E. Therefore the completeness of E provides the convergence of the series. Passing to the limit as $N \to \infty$ in the inequality $\|\sum_0^N x_i\| \leq \sum_0^N \|x_i\|$ and using the continuity of the norm we obtain (5.15).

Exercise 5.3. We consider only the first system $g_k = \sqrt{2/\pi}\,\sin kx$, the proof for the second one being analogous. The fact that (g_k) is orthonormal can be easily checked noticing that g_k are restrictions to $(0, \pi)$ of odd functions, and using the orthogonality of $\sin kx$ in $L^2(-\pi, \pi)$. Analogously, if $f \in L^2(0, \pi)$ let us consider its extension $\tilde{f}$ to $(-\pi, \pi)$ as an odd function and its Fourier series, which obviously contains no cosinus. In $(0, \pi)$ we have

$$\sum_{k=1}^{N} b_k \sin kx = \sum_{k=1}^{N} \langle f, g_k \rangle g_k,$$

where the scalar products are understood in $L^2(0, \pi)$. Therefore, from the convergence of the Fourier series in $L^2(-\pi, \pi)$ to $\tilde{f}$, which implies convergence in $L^2(0, \pi)$ to f, the completeness follows.

Exercise 5.4. Clearly $\langle e_k, e_k \rangle = 1$, while

$$\int_{-\pi}^{\pi} e^{ikx} e^{-ilx}\, dx = \frac{1}{i(k-l)} \left[e^{i(k-l)x}\, dx \right]_{-\pi}^{\pi} = 0 \qquad \text{whenever } k \neq l.$$

As a consequence (e_k) is an orthonormal system.
Since the Fourier series $S_N f = \sum_{-N}^{N} \langle f, e_k \rangle e_k$ of f depends linearly on f, in order to show completeness we need only to show $S_N f \to f$ when f is real-valued and when f is imaginary-valued (i.e. if is real-valued). We consider only the first case, the second one being analogous. Setting $c_k = \langle f, e_k \rangle$, we have

$$c_k = \frac{1}{\sqrt{2\pi}} \int_{\pi}^{\pi} f(x) \cos kx - if(x) \sin kx\, dx.$$

As a consequence, for $k \geq 1$ we have $\sqrt{2/\pi}\, c_k = a_k - ib_k$, where a_k and b_k are the coefficients of the real Fourier series of f, and for $k \leq -1$ we have $\sqrt{2/\pi}\, c_k = a_{-k} + ib_{-k}$. For $k = 0$, instead, we have $\sqrt{2/\pi}\, c_0 = a_0$. Taking into account these relations and setting $b_0 = 0$, we have

$$\sum_{k=-N}^{N} c_k \frac{e^{ikx}}{\sqrt{2\pi}} = \frac{1}{2} \left\{ \sum_{k=1}^{N} (\cos kx + i \sin kx)(a_k - ib_k) \right.$$
$$\left. + \sum_{k=-N}^{-1} (\cos kx + i \sin kx)(a_{-k} - ib_{-k}) \right\}$$
$$= \frac{a_0}{2} + \mathrm{Re}\left(\sum_{k=1}^{N} (\cos kx + i \sin kx)(a_k - ib_k) \right)$$
$$= \frac{a_0}{2} + \sum_{k=1}^{N} a_k \cos kx + b_k \sin kx,$$

and the convergence of $S_N f$ to f follows by the convergence in the real-valued case.

Exercise 5.5. It suffices to note that

$$\frac{1}{2\pi}\left(\int_{-\pi}^{\pi} f(x)e^{-ikx}\,dx\right)^2 = (\langle f, e_k\rangle)^2,$$

where (e_k) is the orthonormal system of Exercise 5.4 and to use its completeness.

Exercise 5.6. From the identity $\sum_{i=0}^{2N} e^{ikz} = (e^{i(2N+1)z} - 1)/(e^{iz} - 1)$, we get

$$\sum_{k=-N}^{N} e^{ikz} = e^{-iNz}\sum_{k=0}^{2N} e^{ikz} = e^{-iNz}\frac{e^{i(2N+1)z} - 1}{e^{iz} - 1} =$$

$$= \frac{e^{i(N+1/2)z} - e^{-i(N+1/2)z}}{e^{iz/2} - e^{-iz/2}} = \frac{\sin((N + 1/2)z)}{\sin(z/2)} \quad (A.15)$$

and we call this term $G_N(z)$. Hence

$$S_N f(x) = \sum_{k=-N}^{N} \frac{1}{2\pi}\left(\int_{-\pi}^{\pi} f(y)e^{-iky}\,dy\right)e^{ikx}$$

$$= \sum_{k=-N}^{N} \frac{1}{2\pi}\int_{-\pi}^{\pi} f(y)e^{ik(x-y)}\,dy$$

$$= \frac{1}{2\pi}\int_{-\pi}^{\pi} f(y)G_N(x - y)\,dy.$$

Using the fact that $\sin((N+1/2)z)/\sin(z/2)$ has, still because of (A.15), mean value 1 on $(-\pi, \pi)$, we get

$$f(x) - S_N f(x) = \frac{1}{2\pi}\int_{-\pi}^{\pi} (f(x) - f(y))G_N(x - y)\,dy.$$

Exercise 5.7. We apply the Parseval identity to the function $f(x) = x^2$, whose Fouries series contains no sinus. It is simple to check, by integration by parts, that $a_0 = 2\pi^2/3$ and that $a_k = 4k^{-2}\cos kx$ for $k \geq 1$. We have then

$$\frac{1}{\pi}\int_{-\pi}^{\pi} x^4\,dx = \frac{2}{5}\pi^4 = \frac{a_0^2}{2} + \sum_{k=1}^{\infty} a_k^2 = \frac{4}{18}\pi^4 + \sum_{k=1}^{\infty}\frac{16}{k^4}.$$

Rearranging terms, we get $\sum_1^{\infty} k^{-4} = \pi^4/90$.

Exercise 5.8. The polynomials P_n are given by $Q_n/\|Q_n\|_2$, where Q_n are recursively defined by $Q_0 = 1$ and

$$Q_n(x) := x^n - \sum_{k=0}^{n-1} \frac{\langle x^n, Q_k \rangle}{\langle Q_k, Q_k \rangle} Q_k(x) = x^n - \sum_{k=0}^{n-1} \langle x^n, P_k \rangle P_k(x) \quad \forall n \geq 1.$$

(a) Since $Q_0 = 1$, $P_0 = 1/\sqrt{2}$ and $Q_1 = x - \langle x, P_0 \rangle P_0 = x$, because $\langle x, P_0 \rangle = 0$. As a consequence $P_1(x) = \sqrt{3/2}x$. Since $\langle x^2, P_1 \rangle = 0$, we have also

$$Q_2(x) = x^2 - \langle x^2, P_0 \rangle P_0 - \langle x^2, P_1 \rangle P_1 = x^2 - \frac{1}{3}$$

and this leads, with simple calculations, to $P_2(x) = \sqrt{45/8}(x^2 - 1/3)$.
(b) Let H be the closure of the vector space spanned by C_n. This space contains all monomials x^n, and therefore all polynomials. Since the polynomials are dense in $C([a, b])$, for the sup norm, they are also dense in $L^2(a, b)$. It follows that $H = L^2(a, b)$. By Proposition 4.13 we conclude that (C_n) is complete.
(c) Set

$$z_n := \sqrt{\frac{2n + 1}{2} \frac{1}{2^n n!}}, \qquad \tilde{P}_n(x) := z_n \frac{d^n}{d^n x}(x^2 - 1)^n$$

Clearly the polynomial $\tilde{P}_n$ has degree n. So, in order to show that $\tilde{P}_n = P_n$, we have to show that $\tilde{P}_n$ is orthogonal to all monomials x^k, $k = 0, \ldots, n-1$, and that $\|\tilde{P}_n\|_2 = 1$. Since $\tilde{P}_n$ has zeros at ± 1 with multiplicity n, all its derivatives at ± 1 with order less than n are zero. Therefore, for $k < n$ we have

$$\langle \tilde{P}_n, x^k \rangle = z_n \left\{ \left[x^k \frac{d^{n-1}}{d^{n-1}x}(x^2 - 1)^n \right]_{-1}^1 - k \int_{-1}^1 x^{k-1} \frac{d^{n-1}}{d^{n-1}x}(x^2 - 1)^n \, dx \right\}$$

$$= \cdots$$

$$= (-1)^k k! z_n \left[\frac{d^{n-k}}{d^{n-k}x}(x^2 - 1)^n \right]_{-1}^1 = 0.$$

In order to prove that $\|\tilde{P}_n\|_2 = 1$, still integrating by parts we have

$$\langle \tilde{P}_n, \tilde{P}_n \rangle = -z_n^2 \int_{-1}^1 \frac{d^{n-1}}{d^{n-1}x}(x^2 - 1)^n \frac{d^{n+1}}{d^{n+1}x}(x^2 - 1)^n \, dx = \cdots$$

$$= z_n^2 \int_{-1}^1 (1 - x^2)^n \frac{d^{2n}}{d^{2n}x}(x^2 - 1)^n \, dx.$$

(A.16)

On the other hand

$$\int_{-1}^{1} (1 - x^2)^n \, dx = 2n \int_{-1}^{1} (1 - x^2)^{n-1} x^2 \, dx$$

$$= -2n \int_{-1}^{1} (1 - x^2)^n \, dx + 2n \int_{-1}^{1} (1 - x^2)^{n-1} \, dx,$$

so that

$$\int_{-1}^{1} (1 - x^2)^n \, dx = \frac{2n}{2n+1} \int_{-1}^{1} (1 - x^2)^{n-1} \, dx = \cdots$$

$$= \frac{(2n)!!}{(2n+1)!!} \int_{-1}^{1} (1 - x^2)^0 \, dx = \frac{2(2n)!!}{(2n+1)!!}.$$

Taking into account that

$$\frac{d^{2n}}{d^{2n}x}(x^2 - 1)^n = (2n)! = (2n)!!(2n-1)!! = 2^n n!(2n-1)!!$$

from (A.16) we get

$$\langle \tilde{P}_n, \tilde{P}_n \rangle = \frac{2n+1}{2} \frac{1}{2^{2n}(n!)^2} \frac{2(2n)!!}{(2n+1)!!} 2^n n!(2n-1)!! = 1.$$

Exercise 5.9. Recall that

$$c_k = \frac{1}{2\pi} \int_{-\pi}^{\pi} f(x) e^{-ikx} \, dx.$$

Integrating by parts once and using that $f(-\pi) = f(\pi)$ we get

$$c_k = \frac{1}{ik} \frac{1}{2\pi} \int_{-\pi}^{\pi} f'(x) e^{-ikx} \, dx.$$

Continuing in this way, in m steps we get

$$c_k = \frac{1}{(ik)^m} \frac{1}{2\pi} \int_{-\pi}^{\pi} f^{(m)}(x) e^{-ikx} \, dx.$$

Chapter 6

Exercise 6.1. Let us prove the inclusion

$$(\mathscr{F}_1 \times \mathscr{F}_2) \times \mathscr{F}_3 \subset \mathscr{F}_1 \times (\mathscr{F}_2 \times \mathscr{F}_3),$$

the proof of the converse one being analogous. We have to show that all products $A \times B$, with $A \in \mathscr{F}_1 \times \mathscr{F}_2$ and $B \in \mathscr{F}_3$ belong to $\mathscr{F}_1 \times (\mathscr{F}_2 \times \mathscr{F}_3)$. Keeping B fixed, the class of sets A for which this property holds is a σ–algebra that contains the π–system of measurable rectangles $A = A_1 \times A_2$ (because $A \times B = A_1 \times (A_2 \times B)$ and $A_2 \times B \in \mathscr{F}_2 \times \mathscr{F}_3$), and therefore the whole product σ–algebra $\mathscr{F}_1 \times \mathscr{F}_2$.

For all A in the product σ–algebra we have

$$
\begin{aligned}
(\mu_1 \times \mu_2) \times \mu_3(A) &= \int_{X_1 \times X_2} \mu_3(A_{x_1 x_2})\, d\mu_1 \times \mu_2(x_1, x_2) \\
&= \int_{X_1} \int_{X_2} \mu_3(A_{x_1 x_2})\, d\mu_2(x_2)\, d\mu_1(x_1) \\
&= \int_{X_1} \mu_2 \times \mu_3(A_{x_1})\, d\mu_1(x_1) = \mu_1 \times (\mu_2 \times \mu_3)(A).
\end{aligned}
$$

Exercise 6.2. Obviously the cubes belong to $\mathsf{X}_1^n \, \mathscr{B}(\mathbb{R})$, and thanks to Lemma 6.9 the same is true for the open sets. It follows that $\mathscr{B}(\mathbb{R}^n)$ is contained in $\mathsf{X}_1^n \, \mathscr{B}(\mathbb{R})$. Let us consider the class

$$
\mathscr{M} := \big\{ B \subset \mathbb{R} : B \times \mathbb{R} \times \cdots \times \mathbb{R} \in \mathscr{B}(\mathbb{R}^n) \big\}.
$$

This class contains the open sets (because the product of open sets is open) and it is a σ–algebra, so it contains $\mathscr{B}(\mathbb{R})$. We have thus proved that all rectangles $B_1 \times \mathbb{R} \times \cdots \times \mathbb{R}$, with B_1 Borel belong to $\mathscr{B}(\mathbb{R}^n)$. By a similar argument we can show that all rectangles

$$
\mathbb{R} \times \cdots \times \mathbb{R} \times B_i \times \mathbb{R} \times \cdots \times \mathbb{R}
$$

are Borel. Intersecting rectangles in these families we obtain that all rectangles with Borel sides belong to $\mathscr{B}(\mathbb{R}^n)$ and we conclude.

Exercise 6.3. Assume that $A, B \in \mathscr{L}_1$; then there exist Borel sets A', B' and Borel Lebesgue negligible sets N_A, N_B with $A \triangle A' \subset N_A$ and $B \triangle B' \subset N_B$. Since $A' \times B' \in \mathscr{B}(\mathbb{R}^2)$, by the previous exercise,

$$
(A \times B) \triangle (A' \times B') \subset (N_A \times \mathbb{R}) \cup (\mathbb{R} \times N_B)
$$

and $N_A \times \mathbb{R}$ and $\mathbb{R} \times N_B$ are $\mathscr{L}^2$ negligible, we obtain that $A \times B \in \mathscr{L}_2$. This proves that $\mathscr{L}_2$ contains the generators of $\mathscr{L}_1 \times \mathscr{L}_1$, and therefore the whole σ–algebra. In order to show the strict inclusion, we consider the set $E = F \times \{0\}$, where $F \subset \mathbb{R}$ is not Lebesgue measurable. Since E is $\mathscr{L}^2$–negligible we have $E \in \mathscr{L}_2$. On the other hand, since the 0

section E^0 coincides with F, and therefore it does not belong to $\mathscr{L}_1$, the set E can't belong to the product of the two σ–algebras.

Exercise 6.4. Let $\mathscr{A}$ be the σ–algebra generated by these sets; since these sets are obviously cylindrical, $\mathscr{A}$ is contained in the product σ–algebra. The class of sets $B \subset \bigtimes_1^n X_i$ such that $B \times X_{n+1} \times X_{n+2} \times \cdots \in \mathscr{A}$ is a σ–algebra containing the measurable rectangles $A_1 \times \cdots \times A_n$, and therefore contains the product σ–algebra $\bigtimes_1^n \mathscr{F}_i$. Therefore $\mathscr{A}$ contains the cylindrical sets and, by definition, the whole product σ–algebra.

Exercise 6.5. The sections $T_y := \{(x, z) \,:\, (x, y, z) \in T\}$ are squares with length side $2\sqrt{r^2 - |y|^2}$ for $0 \le |y| \le r$, hence

$$\mathscr{L}^3(T) = \int_{-r}^{r} \mathscr{L}^2(T_y)\,dy = 8\int_0^r (r^2 - y^2)\,dy = 8\left(r^3 - \frac{1}{3}r^3\right) = \frac{16}{3}r^3.$$

Exercise 6.6. For $x \in \mathbb{R}^n$ (with $n \ge 3$) let

$$r := (x_1^2 + x_2^2)^{1/2}, \qquad A_r := \left\{(x_3, \ldots, x_n) \,:\, (x_3^2 + \cdots + x_n^2) < 1 - r^2\right\}.$$

Then, using polar coordinates we get

$$\omega_n = \int_{\{r<1\}} \mathscr{L}^{n-2}(A_r)\,dx_1 dx_2 = 2\pi\,\omega_{n-2} \int_0^1 r(1 - r^2)^{(n-2)/2}\,dr$$
$$= \frac{2\pi}{n}\omega_{n-2}.$$

Therefore

$$\omega_{2k} = \frac{2^{k-1}\pi^{k-1}}{2k(2k-2)\cdots 4}\,\omega_2 = \frac{\pi^k}{k!}$$

and an analogous argument gives $\omega_{2k+1} = 2^{k+1}\pi^k/(2k+1)!!$.

Exercise 6.7. In order to show that $\omega_n = \dfrac{\pi^{n/2}}{\Gamma(\frac{n}{2}+1)}$ we show that the right hand side satisfies the same recursion formula of the previous exercise. Since (thanks to the identities $\Gamma(1) = 1$, $\Gamma(1/2) = \sqrt{\pi}$) the formula holds when $n = 1,\, 2$, this will prove that the identity holds for all n. For $n \ge 2$ we have

$$\frac{\pi^{n/2}}{\Gamma(\frac{n}{2}+1)} = \frac{\pi \cdot \pi^{(n-2)/2}}{\frac{n}{2}\Gamma(\frac{n}{2})} = \frac{2\pi}{n}\,\frac{\pi^{(n-2)/2}}{\Gamma(\frac{(n-2)}{2}+1)}.$$

Exercise 6.8. We know, by Exercise 2.4, that there exist a λ–negligible set $N \in \mathscr{F} \times \mathscr{G}$ and a $\mathscr{F} \times \mathscr{G}$–measurable function $\tilde{F} : X \times Y \to [0, +\infty]$ such that $\{F \neq \tilde{F}\}$ is contained in N. By applying the Fubini–Tonelli theorem to $\mathbb{1}_N$ we obtain that N_x is ν–negligible in Y for μ–a.e. $x \in X$. Since $\{F(x, \cdot) \neq \tilde{F}(x, \cdot)\} \subset N_x$, still Exercise 2.4 gives that $F(x, \cdot)$ is ν–measurable for μ–a.e. $x \in X$. This proves statement (i). Since, still for μ–a.e. $x \in X$, the integral on Y (with respect to ν) of $F(x, \cdot)$ coincides with the integral of $\tilde{F}(x, \cdot)$, statements (ii) and (iii) follow by applying the Fubini–Tonelli theorem to $\tilde{F}$.

Exercise 6.9. Indeed, $\mu(D_y) = \mu(\{y\}) = 0$ for all $y \in Y$, so that $\int_Y \mu(D_y) \, d\nu(y) = 0$. On the other hand, $\nu(D_x) = \nu(\{x\}) = 1$ for all $x \in X$, so that $\int_X \nu(D_x) \, d\mu(x) = 1$.

Exercise 6.10. Let $(h(k))$ be a subsequence such that $\sum_k \| f_{h(k)} - f \|_1$ is convergent. Then the Fubini–Tonelli theorem gives

$$\int_X \left(\sum_{k=0}^{\infty} \int_Y |f_{h(k)}(x, y) - f(x, y)| \, d\nu(y) \right) d\mu(x)$$

$$= \sum_{k=0}^{\infty} \int_{X \times Y} |f_{h(k)}(x, y) - f(x, y)| \, d\mu \times \nu < \infty.$$

It follows that $\sum_k \| f_{h(k)}(x, \cdot) - f(x, \cdot) \|_{L^1(\nu)}$ is finite for μ–a.e. $x \in X$, and for any such x the functions $f_{h(k)}(x, \cdot)$ converge to f in $L^1(\nu)$. Choosing $Y = \{\bar{y}\}$ and $\nu = \delta_{\bar{y}}$, to provide a counterexample it is sufficient to consider any example (see Remark 3.7) of a sequence converging in L^1 but not μ–almost everywhere.

Exercise 6.11. It suffices to apply (6.15) to $|h|$ to show that $\int |h| \, df\mu$ is finite if and only if $\int |h| f \, d\mu$ is finite.

Exercise 6.12. We prove the property for the sup, the property for the inf being analogous. If $A = B_1 \cup B_2$ with $B_1 \in \mathcal{F}$ and $B_2 \in \mathcal{F}$ disjoint, we have

$$f\mu(B_1) + g\mu(B_2) = \int_{B_1} f \, d\mu + \int_{B_2} g \, d\mu$$

$$\leq \int_{B_1} f \vee g \, d\mu + \int_{B_2} f \vee g \, d\mu$$

$$= \int_A f \vee g \, d\mu.$$

The arbitrariness of this decomposition, proves that $[(f\mu) \vee (g\mu)](A) \leq (f \vee g)\mu(A)$. The converse inequality can be obtained noticing that, in the chain of equalities-inequality above, the inequality becomes an equality if we choose $B_1 = A \cap \{f \geq g\}$ and $B_2 = A \cap \{f > g\}$.

Exercise 6.13. It is easy to check that $\underline{\mu} \leq \mu_i$ (respectively, $\overline{\mu} \geq \mu_i$) for all $i \in I$, and that any measure ν with this property is less than $\underline{\mu}$ (resp.

greater than $\overline{\mu}$): just write $\nu(B) = \sum_k \nu(B_k) \leq \sum_k \mu_{i(k)}(B_k)$ (resp. $\geq \sum_k \mu_{i(k)}(B_k)$. So, it remains to show that $\underline{\mu}$ and $\overline{\mu}$ are σ-additive. For any map $i : \mathbb{N} \to I$, A_1, $A_2 \in \mathscr{F}$ disjoint and any countable $\mathscr{F}$–measurable partition of $A_1 \cup A_2$ we have

$$\sum_{k=0}^{\infty} \mu_{i(k)}(B_k) = \sum_{k=0}^{\infty} \mu_{i(k)}(B_k \cap A_1) + \sum_{k=0}^{\infty} \mu_{i(k)}(B_k \cap A_2).$$

Estimating the right hand side from below with $\underline{\mu}(A_1) + \underline{\mu}(A_2)$ we get (because (B_k) is arbitrary) that $\underline{\mu}$ is superadditive, i.e. $\underline{\mu}(A_1 \cup A_2) \geq \underline{\mu}(A_1) + \underline{\mu}(A_2)$. With a similar argument one can prove not only that $\overline{\mu}$ is subadditive, but also that $\overline{\mu}$ is σ–subadditive (it suffices to consider a countable $\mathscr{F}$–measurable family, instead of 2 sets).

Now, let us prove that $\underline{\mu}$ is subadditive and $\overline{\mu}$ is superadditive. Let A_1, $A_2 \in \mathscr{F}$ be disjoint and let B_k^1, B_k^2 be countable $\mathscr{F}$–measurable partitions of A_1 and A_2 respectively. If i_1, $i_2 : \mathbb{N} \to I$ we define $i(2k) = i_1(k)$, $B_{2k} = B_k^1$ and $i(2n+1) = i_2(n)$, $B_{2k+1} = B_k^2$, so that

$$\underline{\mu}(A_1 \cup A_2) \leq \sum_{k=0}^{\infty} \mu_{i(k)}(B_k) = \sum_{k=0}^{\infty} \mu_{i_1(k)}(B_k^1) + \sum_{k=0}^{\infty} \mu_{i_2(k)}(B_k^2).$$

By the arbitrariness of B_k^1, B_k^2, i_1 and i_2 we conclude that $\underline{\mu}(A_1 \cup A_2) \leq \underline{\mu}(A_1) + \underline{\mu}(A_2)$. With a similar argument one can prove that $\underline{\mu}$ is even σ–subadditive (one has to use a bijection between $\mathbb{N} \times \mathbb{N}$ and $\mathbb{N}$) and that $\overline{\mu}$ is superadditive.

Exercise 6.14. If for all $\varepsilon > 0$ there exists $\delta > 0$ satisfying

$$A \in \mathscr{F}, \ \mu(A) < \delta \qquad \Longrightarrow \qquad \nu(A) < \varepsilon$$

then $\nu \ll \mu$: indeed, if $\mu(A) = 0$ the implication above holds for all $\varepsilon > 0$, hence $\nu(A) = 0$. If ν is finite, to prove the converse we argue by contradiction. Assume that, for some ε_0, we can find sets $A_n \in \mathscr{F}$ with $\mu(A_n) < 2^{-n}$ and $\nu(A_n) \geq \varepsilon_0$. Then, by the Borel–Cantelli lemma the set $A := \limsup_n A_n$ is μ–negligible. On the other hand, we have

$$\nu\left(\bigcup_{m=n}^{\infty} A_m\right) \geq \nu(A_n) \geq \varepsilon_0$$

and therefore (here we use the assumption that ν is finite) $\nu(A) \geq \varepsilon_0$, contradicting the absolute continuity of ν with respect to μ.

Exercise 6.15. Let $B \in \mathscr{F}$ be a μ–negligible set where ν is concentrated. Then $\nu(E) = \nu(E \cap B)$ for all $E \in \mathscr{F}$. But, by the absolute continuity of ν with respect to μ, we have $\nu(E \cap B) = 0$ because $E \cap B \subset B$ is μ–negligible.

Exercise 6.16. Let $B \in \mathscr{F}$ be a ν–negligible set where σ is concentrated. Then

$$\sigma(E) = \sigma(E \cap B) \leq \mu(E \cap B) + \nu(E \cap B) \leq \mu(E) \qquad \forall E \in \mathscr{F},$$

where we used the fact that $\nu(E \cap B) = 0$ because $E \cap B \subset B$ is ν–negligible.

Exercise 6.17. It is easy to check that the class of functions f satisfying $f\mu \leq \nu$ is a lattice. Hence, given a maximizing sequence (f_h) in (6.20), possibly replacing f_h by $\max_{i \leq n} f_i$, we can assume that $f_h \uparrow f$. The monotone convergence theorem gives that f is a maximizer.

In order to show that $\nu = f\mu$ we set $\sigma = \nu - f\mu \geq 0$ and notice that σ satisfies the following property:

$$t > 0, \quad B \in \mathscr{F}, \quad t\mathbb{1}_B\mu \leq \sigma \qquad \Longrightarrow \qquad \mu(B) = 0. \qquad \text{(A.17)}$$

Indeed, the integrals $\int_X (f + t\mathbb{1}_B)\, d\mu$ and $\int_X f\, d\mu$ have to coincide, because $(f + t\chi_B)\mu \leq \nu$.

Exercise 6.18. We have to prove that any measure σ satisfying (A.17) is concentrated on a μ-negligible set. To this aim, let us consider the problem

$$\inf\{\mu(A) : \; A \in \mathscr{F}, \; \sigma \text{ is concentrated on } A\}.$$

By taking the intersection of a minimizing sequence it is easy to check that also this problem has a solution A; we have to show that $\mu(A) = 0$. By the minimality of A, the implication

$$\mathscr{F} \ni B \subset A, \quad \mu(B) > 0 \qquad \Longrightarrow \qquad \sigma(B) > 0 \qquad \text{(A.18)}$$

holds. Let us consider the numbers

$$\xi_h := \sup\left\{\mu(B) : \; \mathscr{F} \ni B \subset A, \; \chi_B\mu \geq 2^h\mathbb{1}_B\sigma\right\}$$

and let us prove that $\xi_h \to 0$ as $h \to \infty$. Given maximizers $B_h \subset A$, whose existence is easy to check, we have $\mu(B_h) \geq 2^h\sigma(B_h)$ and in particular $\sum_h \sigma(B_h) < \infty$. Hence

$$\sigma\left(\limsup_{h \to \infty} B_h\right) = 0$$

and (A.18) tells us that necessarily

$$0 = \mu\left(\limsup_{h\to\infty} B_h\right) \geq \limsup_{h\to\infty} \mu(B_h).$$

Let us show now that the maximality of B_h implies that $\mu(C) \leq 2^h\sigma(C)$ for any set $C \subset A \setminus B_h$, i.e. $t\mathbb{1}_{A\setminus B_h}\mu \leq \sigma$. Indeed, if there is $C_0 \subset A \setminus B_h$ with $\mu(C_0) > 2^h\sigma(C_0)$, the maximality of B_h provides a minimal integer $h_1 \geq 1$ and $C_1 \subset C_0$ satisfying $\mu(C_1) \leq 2^h\sigma(C_1) - 1/h_1$. Let us consider $C_0 \setminus C_1$; we still have $\mu(C_0 \setminus C_1) > 2^h\sigma(C_0 \setminus C_1)$ and the maximality of B_h provide a minimal integer $h_2 \geq h_1$ and $C_2 \subset C_0 \setminus C_1$ satisfying $\mu(C_2) \leq 2^h\sigma(C_2) - 1/h_2$. Continuing in this way we have a nondecreasing sequence (h_i) of integers and $(C_i) \subset \mathscr{F}$ such that $\mu(C_i) \leq 2^h\sigma(C_i) - 1/h_i$ and $C_i \subset C_0 \setminus \cup_{j=1}^{i-1}C_j$ for all $i \geq 2$; moreover h_i is the least integer for which there is such C_i. Now $\lim_i h_i = \infty$, since the C_i are pairwise disjoint. Setting $C = C_0 \setminus \cup_1^\infty C_i$, for all $F \in \mathscr{F}$ contained in C, since $F \subset C_0 \setminus \cup_1^{i-1}C_j$ for all $i \geq 2$, we have $\mu(F) \geq 2^h\sigma(F) - 1/(h_i - 1)$ (if $h_i \geq 2$) and then $\mu(F) \geq 2^h\sigma(F)$. Hence $B_h \cup C$ is an admissible set for the maximum problem defining ξ_h, against the maximality of B_h.

We choose h in such a way that $\xi_h < \mu(A)$ and set $t = 2^{-h}$, $B = A \setminus B_h$ in (A.17). From (A.17) we conclude that $\mu(B) = 0$, contradicting the fact that $\mu(B) = \mu(A) - \xi_h > 0$.

Exercise 6.19. Let $\nu = \nu^+ - \nu^-$ and let $\nu^+ = \nu_a^+ + \nu_s^+$, $\nu^- = \nu_a^- + \nu_s^-$ be the Lebesgue decompositions with respect to μ of ν^+ and ν^- respectively. Then, $\nu_a := \nu_a^+ - \nu_a^-$ and $\nu_s := \nu_s^+ - \nu_s^-$ provide a decomposition $\nu = \nu_a + \nu_s$ with ν_a, ν_s signed, $|\nu_a| \ll \mu$ and $|\nu_s| \perp \mu$.

If μ is signed and A provides a Hahn decomposition of μ (i.e. $\mu^+(E) = \mu(E \cap A)$ and $\mu^-(E) = -\mu(E \cap A^c)$), we repeat the decomposition above in A, relative to ν and μ^+, and in $B = A^c$, relative to ν and μ^-. Denoting by $\nu_a^A + \nu_s^A$ and $\nu_a^B + \nu_s^B$ the two decompositions obtained,

$$\nu_a(E) := \nu_a^A(E \cap A) + \nu_a^B(E \cap B), \qquad \nu_s(E) := \nu_s^A(E \cap A) + \nu_s^B(E \cap B)$$

provides the desired decomposition $\nu = \nu_a + \nu_s$ with $|\nu_a| \ll |\mu|$ and $|\nu_s| \perp |\mu|$.

The uniqueness of these decompositions can be proved with the same argument used in the case of nonnegative measures.

Exercise 6.20. Let $B \in \mathscr{F}$ and let (B_i) be a $\mathscr{F}$–measurable partition of B; since

$$\sum_{i=0}^{\infty} |f\mu(B_i)| = \sum_{i=0}^{\infty}\left|\int_{B_i} f\,d\mu\right| \leq \sum_{i=0}^{\infty}\int_{B_i} |f|\,d\mu = \int_B |f|\,d\mu,$$

we obtain that $|f\mu|(B) \leq |f|\mu(B)$. To prove the converse inequality fix $\varepsilon > 0$ and define $B_i = B \cap f^{-1}(I_i)$, where $I_i = \varepsilon[i, i+1)$, $i \in \mathbb{Z}$. Since the oscillation of $|f - \varepsilon i|$ and $||f| - \varepsilon|i||$ in $f^{-i}(I_i)$ are less than ϵ, we get

$$\left|\int_{B_i} f\, d\mu - \varepsilon i\, \mu(B_i)\right| \leq \varepsilon\mu(B_i), \qquad \left|\int_{B_i} |f|\, d\mu - \varepsilon|i|\mu(B_i)\right| \leq \varepsilon\mu(B_i),$$

hence

$$\left|\int_{B_i} |f|\, d\mu - \left|\int_{B_i} f\, d\mu\right|\right| \leq 2\varepsilon\mu(B_i).$$

It follows that

$$\sum_{i\in\mathbb{Z}} |f\mu(B_i)| = \sum_{i\in\mathbb{Z}} \left|\int_{B_i} f\, d\mu\right| \geq \sum_{i\in\mathbb{Z}} \int_{B_i} |f|\, d\mu - 2\varepsilon\mu(B_i)$$

$$= \int_B |f|\, d\mu - 2\varepsilon\mu(B).$$

Since ε is arbitrary the converse inequality follows.

Exercise 6.21. If $x < 0$ or $x \geq 1$ all repartition functions are respectively equal to 0 or 1, so we need to consider only the case $x \in [0, 1)$. The repartition function of $\mathbb{1}_{[0,1]}\mathscr{L}^1$ obviously is equal to x, while

$$\mu_h((-\infty, x]) = \frac{\#\{i \in [1, h] \,:\, i \leq hx\}}{h} = \frac{[hx]}{h},$$

where $[s]$ denotes the integer part of s. Using the inequalities $s - 1 < [s] \leq s$ with $s = hx$ we obtain that $\mu_h((-\infty, x]) \to x$.

Exercise 6.22. The argument is similar to the one used in the proof of Theorem 6.27: if $y < x < y'$ and $y, y' \in D$ we have

$$F(y) = \lim_{h\to\infty} F_h(y) \leq \liminf_{h\to\infty} F_h(x) \leq \limsup_{h\to\infty} F_h(x)$$

$$\leq \lim_{h\to\infty} F_h(y') = F(y').$$

Letting $y \uparrow x$ and $y' \downarrow x$, we conclude.

Exercise 6.23. We define $a_{-h^2} = \mu((-\infty, -h])$ and, for $-h^2 < i \leq h^2$, $a_i = \mu((i-1)/h, i/h])$. Let us denote by μ_h the measure obtained in this way. If $x \in (-h, h]$ and i is the smallest integer in $(-h^2, h^2]$ such that $x \leq i/h$, we have

$$\mu\left(\left(-\infty, x - \frac{1}{h}\right]\right) \leq \mu\left(\left(-\infty, \frac{i-1}{h}\right]\right) = \sum_{j=-h^2}^{i-1} a_i \leq \mu_h((-\infty, x]).$$

If x is not an atom of μ, this proves that

$$\liminf_{h} \mu_h((-\infty, x]) \geq \mu((-\infty, x]).$$

Analogously

$$\mu\left(\left(-\infty, x + \frac{1}{h}\right]\right) \geq \mu\left(\left(-\infty, \frac{i}{h}\right]\right) = \sum_{j=-h^2}^{i} a_i \geq \mu_h((-\infty, x]).$$

If x is not an atom of μ, this proves that

$$\limsup_{h} \mu_h((-\infty, x]) \leq \mu((-\infty, x]).$$

Exercise 6.24. Let us assume that (6.31) holds. If $F_i(x) \to 1$ as $x \to +\infty$ uniformly in $i \in I$, for any $\varepsilon > 0$ we can find x such that $1 - F_i(x) < \varepsilon/2$ for all $i \in I$. Analogously, we can find $y < x$ such that $F_i(y) < \varepsilon/2$ for all $i \in I$. Then, the interval $I = (y, x]$ satisfies $\mu_i(I) > 1 - \varepsilon$ for all $i \in I$, because $I^c = (-\infty, y] \cup (x, +\infty)$.

Exercise 6.25. If μ is the weak limit and $\varepsilon > 0$ is given, let us choose an integer $n \geq 1$ such that $\mu([1 - n, n - 1]) > 1 - \varepsilon$ and points $x \in (-n, 1 - n)$ and $y \in (n - 1, n)$ where the repartition functions of μ_h are converging to the repartition function of μ. Then, since $\mu((\infty, x]) + 1 - \mu((-\infty, y]) = \mu(\mathbb{R} \setminus (x, y)) < \varepsilon$, there exists $n_\varepsilon \in \mathbb{N}$ such that $\sup_{n \geq n_\varepsilon} \mu_n((\infty, x]) + 1 - \mu_n((-\infty, y]) < \varepsilon$. Let now x' and y' be satisfying

$$\mu_n((\infty, x']) + 1 - \mu_n((-\infty, y']) < \varepsilon \qquad \forall n = 0, \ldots, n_\varepsilon - 1.$$

Then, the interval $I = [\min\{x, x'\}, \max\{y, y'\}]$ satisfies $\inf_n \mu_n(I) > 1 - \varepsilon$.

Exercise 6.26.

(a) $\lim_h \int_{\mathbb{R}} g \, d\mu_h = \int_{\mathbb{R}} g \, d\mu \qquad \forall g \in C_b(\mathbb{R})$ (that is, (6.32));
(b) $\lim_h \int_{\mathbb{R}} g \, d\mu_h = \int_{\mathbb{R}} g \, d\mu \qquad \forall g \in C_c(\mathbb{R})$;
(c) F_h converge to F on all points where F is continuous;
(d) F_h converge to F on a dense subset of $\mathbb{R}$;
(e) $\lim_h \mu_h(\mathbb{R}) = \mu(\mathbb{R})$;
(f) (μ_h) is tight.

We consider the functions $\rho_h(x) := \rho(x + h)$, where $\rho(x) = (2\pi)^{-\frac{1}{2}} e^{-\frac{x^2}{2}}$ is the Gaussian, and $\mu_h = \rho_h \lambda$ (λ being the Lebesgue measure), $\mu = 0$.

In this case (c), (d), do not hold, because $F_h(x) \to 1 \neq 0 = F(x)$ for all $x \in \mathbb{R}$, (e) does not hold and (b) holds.

$a \Rightarrow b, e$. This is easy, because $C_c(\mathbb{R}) \subset C_b(\mathbb{R})$ and $\mathbb{1}_{\mathbb{R}} \in C_b(\mathbb{R})$.

$a \Rightarrow c$. This follows by second part of the proof of Theorem 6.28.

$d \Leftrightarrow c$. This is Exercise 6.22.

$b \wedge e \Rightarrow c$. This follows by the same argument used in the proof of second part of Theorem 6.28: the sequence (g_k) monotonically convergent to $\mathbb{1}_A$ can be chosen in $C_c(\mathbb{R})$, and this shows that $\liminf_h \mu_h(A) \geq \mu(A)$ for all $A \subset \mathbb{R}$ open. Using (e) and passing to the complementary sets, we obtain $\limsup_h \mu_h(C) \leq \mu(C)$ for all $C \subset \mathbb{R}$ closed.

$d \Rightarrow f$. This follows by the same argument used in the solution of Exercise 6.25.

$d \wedge f \Rightarrow e$. For all $x \in D$, with D dense, we have $\lim_h \mu_h((-\infty, x]) = \mu((-\infty, x])$. Since $\mu_h((-\infty, x]) \to \mu_h(\mathbb{R})$ as $x \to +\infty$ uniformly in h, we can pass to the limit as $x \in D \to +\infty$ to obtain $\lim_h \mu_h(\mathbb{R}) = \lim_{x \to +\infty} \mu((-\infty, x]) = \mu(\mathbb{R})$.

$d \wedge f \Rightarrow a$. This follows by the same argument used in the first part of the proof of Theorem 6.28, choosing the points t_i in the partitions to be in the dense set where convergence occurs.

Exercise 6.27. Set

$$g(\xi) := \frac{1}{\sqrt{2\pi\sigma^2}} \int_{\mathbb{R}} e^{i\xi x} e^{-x^2/(2\sigma^2)} \, dx.$$

Notice that $g(0) = 1$, and that differentiation theorems under the integral sign [2] and an integration by parts give

$$g'(\xi) = \frac{1}{\sqrt{2\pi\sigma^2}} \int_{\mathbb{R}} i e^{i\xi x} (x e^{-x^2/(2\sigma^2)}) \, dx$$

$$= \frac{\sigma^2}{\sqrt{2\pi\sigma^2}} \int_{\mathbb{R}} i \frac{d}{dx} e^{i\xi x} e^{-x^2/(2\sigma^2)} \, dx$$

$$= -\frac{\xi\sigma^2}{\sqrt{2\pi\sigma^2}} \int_{\mathbb{R}} e^{i\xi x} e^{-x^2/(2\sigma^2)} \, dx.$$

Therefore g satisfies the linear differential equation $g'(\xi) = -\sigma^2\xi g(\xi)$, whose general solution is $g(\xi) = c e^{-\sigma^2\xi^2/2}$. Taking into account that $g(0) = 1$, $c = 1$.

---

[2] In this case, the application of the theorem is justified by the fact that $\sup_{\xi \in I} |\frac{d}{d\xi} e^{i\xi x} e^{-x^2/(2\sigma^2)}|$ is Lebesgue integrable for all bounded intervals I

Exercise 6.28. Let us approximate μ by $\mu_n = \mathbb{1}_{(-n,n)}\mu$; using the inequality

$$|e^{i\xi x} - e^{i\eta x}| \le |x||\xi - \eta| \qquad x, \xi, \eta \in \mathbb{R}$$

we obtain that

$$|\hat{\mu}_n(\xi) - \hat{\mu}_n(\eta)| \le |\xi - \eta| \int_{\mathbb{R}} |x|\, d\mu_n(x) \le n|\xi - \eta|,$$

therefore $\hat{\mu}_n$ is uniformly continuous. Since $|\hat{\mu}_n(\xi) - \hat{\mu}(\xi)| \le \mu(\mathbb{R} \setminus [-n, n])$, we have that $\hat{\mu}_n \to \hat{\mu}$ uniformly as $n \to \infty$, therefore $\hat{\mu}$ is uniformly continuous (indeed, given $\varepsilon > 0$, find n such that $\sup|\hat{\mu}_n - \hat{\mu}| < \varepsilon/2$ and $\delta = \varepsilon/(2n)$ to obtain $|\hat{\mu}_n(\xi) - \hat{\mu}_n(\eta)| \le \varepsilon/2$ whenever $|\xi - \eta| < \delta$, and then $|\hat{\mu}(\xi) - \hat{\mu}(\eta)| < \varepsilon$).

Exercise 6.29. Obviously $|\hat{\mu}(\xi_0)| = 1$, and we set $c = \hat{\mu}(\xi_0) = e^{i\theta}$ for some $\theta \in \mathbb{R}$. Since

$$\int_{\mathbb{R}} |1 - \bar{c}e^{ix\xi_0}|^2\, d\mu(x) = 2 - \bar{c}c - c\bar{c} = 0,$$

we obtain that $e^{ix\xi_0} = c$ for μ–a.e. $x \in \mathbb{R}$. This implies that $x\xi_0 - \theta \in 2\pi\mathbb{Z}$ for μ–a.e. $x \in \mathbb{R}$, so that μ is concentrated on the set of points $\{(2n\pi + \theta)/\xi_0\}_{n\in\mathbb{N}}$, and it suffices to set $x_0 = \theta/\xi_0$ to obtain the stated representation of μ as a sum of Dirac masses.

Obviously $|\hat{\mu}| \equiv 1$ if μ is a Dirac mass. Conversely, if $|\hat{\mu}| \equiv 1$, we find x_0 with $\mu(\{x_0\}) > 0$ and ξ_0, $\xi_0' \in \mathbb{R} \setminus \{0\}$ with $\xi_0/\xi_0' \notin \mathbb{Q}$ to obtain that μ is concentrated on the set $\{2n\pi/\xi_0 + x_0\}_{n\in\mathbb{N}}$ and on the set $\{2n\pi/\xi_0' + x_0\}_{n\in\mathbb{N}}$. By our choice of ξ_0 and ξ_0', the intersection of the two sets is the singleton $\{x_0\}$, and this proves that $\mu = \delta_{x_0}$.

Chapter 7

Exercise 7.1. Let $C > 0$ be such that $|H(x) - H(y)| \le C|x - y|$ for all $x, y \in \mathbb{R}$. Let $\varepsilon > 0$ and let $\delta > 0$ be such that $\sum_i |f(b_i) - f(a_i)| < \varepsilon/C$ whenever $\sum_i (b_i - a_i) < \delta$. We have $\sum_i |H(f(b_i)) - H(f(a_i))| \le C \sum_i |f(b_i) - f(a_i)|$ whenever $\sum_i (b_i - a_i) < \delta$. In particular, choosing $f(t) = t$, we see that Lipschitz functions are absolutely continuous.

Exercise 7.2. We assume that both $\mathscr{L}^1(E) > 0$ and $\mathscr{L}^1(\mathbb{R} \setminus E) > 0$. Let $a \in \mathbb{R}$ be such that $\mathscr{L}^1((a, \infty) \cap E) > 0$ and $\mathscr{L}^1((a, \infty) \setminus E) > 0$, and define $F(t) = \mathscr{L}^1(E \cap (a, t))$. By our choice of a, $F(t)$ and $(t - a) - F(t)$ are not identically 0 in $(a, +\infty)$.

If $t > a$ is a rarefaction point of E, we have

$$F_+'(t) = \lim_{h\downarrow 0} \frac{F(t + h) - F(t)}{h} = \lim_{h\downarrow 0} \frac{\mathscr{L}^1((t, t + h) \cap E)}{h} = 0.$$

Analogously, $F'_-(t) = 0$ and we find that F' is equal to 0 at all rarefaction points. A similar argument proves that $F' = 1$ at all density points. Let now $t_0 \in (a, \infty)$ where $0 < F(t_0) < (t_0 - a)$ and apply the mean value theorem to obtain $t'_0 \in (a, t_0)$ such that

$$F(t_0) = (t_0 - a)F'(t'_0).$$

By our choice of t_0 it follows that $F'(t'_0) \in (0, 1)$, a contradiction (because either t'_0 is a density point or a rarefaction point).

Exercise 7.3. Assume first that φ is continuous and bounded. Let $H(z) := \int_{f(a)}^{z} \varphi(y)\, dy$. By the (classical) fundamental theorem of the integral calculus, H is differentiable and $H'(z) = \varphi(z)$ for all $z \in f(I)$. By the chain rule and Exercise 7.1, the function

$$F(t) := \int_{f(a)}^{f(t)} \varphi(y)\, dy = H(f(t))$$

is absolutely continuous and it has derivative equal to $H'(f(t))f'(t) = \varphi(f(t))f'(t)$ at all points t where f is differentiable. On the other hand, still by the fundamental theorem of the integral calculus, the function

$$G(t) := \int_{a}^{t} \varphi(f(x))f'(x)\, dx$$

has derivative equal to $(\varphi \circ f)f'$ $\mathscr{L}^1$–a.e. in $[a, b]$. Since both F and G vanish at $t = a$, they coincide.

By the dominated convergence theorem, the identity of the two functions persists if $\varphi = \mathbb{1}_A$, with A open (because $\mathbb{1}_A$ is the pointwise limit of continuous functions). By applying Dynkin's theorem to the class $\mathscr{M}$ of the sets $E \in \mathscr{B}(f(I))$ such that $\int_{f(a)}^{f(t)} \mathbb{1}_E(y)\, dy = \int_{a}^{t} \mathbb{1}_E(f(x))f'(x)\, dx$ we obtain that the formula holds for all $\varphi = \mathbb{1}_E$ with E Borel. Eventually we obtain it for simple functions and, by uniform approximation, for bounded Borel functions.

Exercise 7.4. Choosing $g = \mathbb{1}_N$, by Exercise 7.3 we get $\int_{a}^{b} \mathbb{1}_{f^{-1}(N)} f'\, dx = 0$, because $\mathbb{1}_N \circ f = \mathbb{1}_{f^{-1}(N)}$. Let h^+ and h^- be respectively the positive and negative part of $f'\mathbb{1}_{f^{-1}(N)}$. Since

$$\int_{a}^{b} h^+\, dx - \int_{a}^{b} h^-\, dx = \int_{a}^{b} f'\mathbb{1}_{f^{-1}(N)}\, dx = 0$$

for all intervals (a, b), it follows that $h^+ = h^-$ $\mathscr{L}^1$–a.e. in $\mathbb{R}$. As a consequence, $f' = 0$ $\mathscr{L}^1$–a.e. in $f^{-1}(N)$.

Chapter 8

Exercise 8.1. Both are measures in $(Z, \mathcal{H})$. If $B \in \mathcal{H}$ then $g \circ f_\# \mu(B) = \mu(f^{-1}(g^{-1}(B)))$, because $(g \circ f)^{-1} = f^{-1} \circ g^{-1}$. On the other hand,

$$g_\#(f_\# \mu)(B) = f_\# \mu(g^{-1}(B)) = \mu(f^{-1}(g^{-1}(B))).$$

Exercise 8.2. Let $n \geq 1$ integer, $0 \leq k < 2^n$ and let us consider the interval $I = [k/2^n, (k+1)/2^n)$. Then, $f^{-1}(I)$ is the cylindrical set of all binary sequences $a_0 a_1 \cdots$ such that $a_0 \cdots a_{n-1}$ is the binary expression of k. It follows that

$$\underset{i=0}{\overset{\infty}{\times}} \left(\frac{1}{2}\delta_0 + \frac{1}{2}\delta_1 \right) \left(f^{-1}(I) \right) = \mathscr{L}^1(I).$$

because their common value is 2^{-n}. On the other hand, $f^{-1}(\{1\})$ consists of a single point and therefore the identity above holds for $I = \{1\}$, the common value being 0. By additivity the identity holds for finite unions of sets of this type, a family stable under finite intersections. By the coincidence criterion the two measures coincide.

Exercise 8.3. Let $A \subset \mathbb{R}$ be a dense open set whose complement C has strictly positive Lebesgue measure (Exercise 1.9), and let

$$\varphi(t) := \min\{1, \operatorname{dist}(t, C)\} \qquad t \in \mathbb{R}.$$

By construction the function φ is continuous, nonnegative, bounded by 1, and vanishes precisely on C. Then, set

$$F(t) := \begin{cases} \displaystyle\int_0^t \varphi(s)\,ds & \text{if } t \geq 0; \\ \displaystyle-\int_t^0 \varphi(s)\,ds & \text{if } t < 0. \end{cases}$$

We have $F' = \varphi$, so that $F \in C^1$ and its critical set $C_F = C$ has positive Lebesgue measure. It follows that $F_\# \mathscr{L}^1$ is not absolutely continuous with respect to $\mathscr{L}^1$. Finally, since $\int_a^b \varphi\,dt > 0$ whenever $a < b$ (because $A \cap (a, b) \neq \varnothing$) we obtain that F is strictly increasing.

Exercise 8.4. Recall that $F(C_F)$ is always Lebesgue negligible, regardless of any injectivity assumption on U. Hence, possibly replacing U by $U \backslash C_F$ we can assume with no loss of generality that $C_F = \varnothing$, i.e. $DF(x)$ is nonsingular at any $x \in U$. Recall that, according to the local invertibility theorem, for any $x \in U$ there exists a ball $B_r(x)$ contained in U such that the restriction to F is injective. Now, following the strategy of

Lemma 6.9 we can cover U by a sequence of right open cubes $\{Q_i\}_{i\in I}$, pairwise disjoint, such that the restriction of F to a neighbourhood of Q_i is injective (we keep dividing a cube until this property is achieved). Let $Q_i = \underset{i=1}{\overset{n}{\times}}[a_i, a_i + \delta)$; for $b_i < a_i$ sufficiently close to a_i and $\tilde{Q}_i = \underset{i=1}{\overset{n}{\times}}(b_i, b_i + \delta)$ we have (by injectivity of F on $\tilde{Q}_i$)

$$F_\#(\mathbb{1}_{\tilde{Q}_i}\mathcal{L}^n) = \frac{1}{|J_F| \circ F^{-1}}\mathbb{1}_{F(\tilde{Q}_i)}\mathcal{L}^n$$

and therefore we can pass to the limit to get

$$F_\#(\mathbb{1}_{Q_i}\mathcal{L}^n) = \frac{1}{|J_F| \circ F^{-1}(y)}\mathbb{1}_{F(Q_i)}\mathcal{L}^n.$$

If we add both sides with respect to $i \in I$ we get

$$F_\#(\mathbb{1}_U\mathcal{L}^n) = \sum_{i\in I}\frac{1}{|J_F| \circ F^{-1}(y)}\mathbb{1}_{F(Q_i)}\mathcal{L}^n = \sum_{x\in F^{-1}(y)}\frac{1}{|J_F|(x)}\mathbb{1}_{F(U)}\mathcal{L}^n.$$

References

[1] L. CARLESON, *On the convergence and growth of Fourier series*, Acta Math. **116** (1966), 135–157.

[2] W. F. EBERLEIN, *Notes on Integration I: The Underlying Convergence Theorem*, Comm. Pure Appl. Math. **X** (1957), 357–360.

[3] H. FEDERER, "Geometric Measure Theory", Springer, 1969.

[4] F. RIESZ and B. NAGY, "Functional Analysis", Dover, 1990.

[5] W. RUDIN, "Real and Complex Analysis", McGraw-Hill, 1987.

[6] S. WAGON, "The Banach-Tarski Paradox", Cambridge University Press, 1985.

[7] K. YOSIDA, "Functional Analysis", Springer, 1980.

LECTURE NOTES

This series publishes polished notes dealing with topics of current research and originating from lectures and seminars held at the Scuola Normale Superiore in Pisa.

Published volumes

1. M. Tosi, P. Vignolo, *Statistical Mechanics and the Physics of Fluids*, 2005 (second edition). ISBN 978-88-7642-144-0
2. M. Giaquinta, L. Martinazzi, *An Introduction to the Regularity Theory for Elliptic Systems, Harmonic Maps and Minimal Graphs*, 2005. ISBN 978-88-7642-168-8
3. G. Della Sala, A. Saracco, A. Simioniuc, G. Tomassini, *Lectures on Complex Analysis and Analytic Geometry*, 2006. ISBN 978-88-7642-199-8
4. M. Polini, M. Tosi, *Many-Body Physics in Condensed Matter Systems*, 2006. ISBN 978-88-7642-192-0

   P. Azzurri, *Problemi di Meccanica*, 2007. ISBN 978-88-7642-223-2
5. R. Barbieri, *Lectures on the ElectroWeak Interactions*, 2007. ISBN 978-88-7642-311-6
6. G. Da Prato, *Introduction to Stochastic Analysis and Malliavin Calculus*, 2007. ISBN 978-88-7642-313-0

   P. Azzurri, *Problemi di meccanica*, 2008 (second edition). ISBN 978-88-7642-317-8

   A. C. G. Mennucci, S. K. Mitter, *Probabilità e informazione*, 2008 (second edition). ISBN 978-88-7642-324-6
7. G. Da Prato, *Introduction to Stochastic Analysis and Malliavin Calculus*, 2008 (second edition). ISBN 978-88-7642-337-6
8. U. Zannier, *Lecture Notes on Diophantine Analysis*, 2009. ISBN 978-88-7642-341-3
9. A. Lunardi, *Interpolation Theory*, 2009 (second edition). ISBN 978-88-7642-342-0

10. L. AMBROSIO, G. DA PRATO, A. MENNUCCI, *Introduction to Measure Theory and Integration*, 2011.
ISBN 978-88-7642-385-7, e-ISBN: 978-88-7642-386-4

Volumes published earlier

G. DA PRATO, *Introduction to Differential Stochastic Equations*, 1995 (second edition 1998). ISBN 978-88-7642-259-1

L. AMBROSIO, *Corso introduttivo alla Teoria Geometrica della Misura ed alle Superfici Minime*, 1996 (reprint 2000).

E. VESENTINI, *Introduction to Continuous Semigroups*, 1996 (second edition 2002). ISBN 978-88-7642-258-4

C. PETRONIO, *A Theorem of Eliashberg and Thurston on Foliations and Contact Structures*, 1997. ISBN 978-88-7642-286-7

Quantum cohomology at the Mittag-Leffler Institute, a cura di Paolo Aluffi, 1998. ISBN 978-88-7642-257-7

G. BINI, C. DE CONCINI, M. POLITO, C. PROCESI, *On the Work of Givental Relative to Mirror Symmetry*, 1998. ISBN 978-88-7642-240-9

H. PHAM, *Imperfections de Marchés et Méthodes d'Evaluation et Couverture d'Options*, 1998. ISBN 978-88-7642-291-1

H. CLEMENS, *Introduction to Hodge Theory*, 1998. ISBN 978-88-7642-268-3

Seminari di Geometria Algebrica 1998-1999, 1999.

A. LUNARDI, *Interpolation Theory*, 1999. ISBN 978-88-7642-296-6

R. SCOGNAMILLO, *Rappresentazioni dei gruppi finiti e loro caratteri*, 1999.

S. RODRIGUEZ, *Symmetry in Physics*, 1999. ISBN 978-88-7642-254-6

F. STROCCHI, *Symmetry Breaking in Classical Systems*, 1999 (2000). ISBN 978-88-7642-262-1

L. AMBROSIO, P. TILLI, *Selected Topics on "Analysis in Metric Spaces"*, 2000. ISBN 978-88-7642-265-2

A. C. G. MENNUCCI, S. K. MITTER, *Probabilità ed Informazione*, 2000.

S. V. BULANOV, *Lectures on Nonlinear Physics*, 2000 (2001). ISBN 978-88-7642-267-6

Lectures on Analysis in Metric Spaces, a cura di Luigi Ambrosio e Francesco Serra Cassano, 2000 (2001). ISBN 978-88-7642-255-3

L. CIOTTI, *Lectures Notes on Stellar Dynamics*, 2000 (2001). ISBN 978-88-7642-266-9

S. RODRIGUEZ, *The Scattering of Light by Matter*, 2001. ISBN 978-88-7642-298-0

G. DA PRATO, *An Introduction to Infinite Dimensional Analysis*, 2001. ISBN 978-88-7642-309-3

S. SUCCI, *An Introduction to Computational Physics: – Part I: Grid Methods*, 2002. ISBN 978-88-7642-263-8

D. BUCUR, G. BUTTAZZO, *Variational Methods in Some Shape Optimization Problems*, 2002. ISBN 978-88-7642-297-3

A. MINGUZZI, M. TOSI, *Introduction to the Theory of Many-Body Systems*, 2002.

S. SUCCI, *An Introduction to Computational Physics: – Part II: Particle Methods*, 2003. ISBN 978-88-7642-264-5

A. MINGUZZI, S. SUCCI, F. TOSCHI, M. TOSI, P. VIGNOLO, *Numerical Methods for Atomic Quantum Gases*, 2004. ISBN 978-88-7642-130-0

Made in the USA
Monee, IL
07 July 2026

56549691R00114